T0215344

FEMINISM AND PHILOSOPHY
OF SCIENCE

Feminist perspectives have been increasingly influential on philosophy of science. *Feminism and Philosophy of Science* is designed to introduce the newcomer to the central themes, issues and arguments of this burgeoning area of study.

Elizabeth Potter engages in a rigorous and well-organized study that takes in the views of key feminist theorists – Nelson, Wylie, Anderson, Longino and Harding – whose arguments exemplify contemporary feminist philosophy of science. The book is divided into six chapters looking at important themes:

- naturalized feminist empiricism
- feminist value theory
- feminist conceptual empiricism
- standpoint epistemologies of science
- value-free science

Arranged thematically, *Feminism and Philosophy of Science* looks at the spectrum of views that have arisen in the debate, and unpacks the arguments on key topics such as value-free science, values, objectivity, standpoint and relativism. It assumes no previous knowledge of the subject, and is written in an accessible, student-friendly style. It will be an important source for students of philosophy, philosophy of science, gender studies and feminist studies.

Elizabeth Potter is Alice Andrew Quigley Professor of Women's Studies and Program Head at Mills College, Oakland, USA.

UNDERSTANDING FEMINIST PHILOSOPHY
Edited by Linda Martin Alcoff

This major new series is designed for students who have typically completed an introductory course in philosophy and are coming to feminist philosophy for the first time. Each book clearly introduces a core undergraduate subject in philosophy, from a feminist perspective, examining the role gender plays in shaping our understanding of philosophy and related disciplines. Each book offers students an accessible transition to higher-level work on that subject and is clearly written, by an experienced author and teacher, with the beginning student in mind.

GENDER AND AESTHETICS
Carolyn Korsmeyer

FEMINISM AND MODERN PHILOSOPHY
Andrea Nye

FEMINISM AND EPISTEMOLOGY
Phyllis Rooney

FEMINISM AND PHILOSOPHY OF SCIENCE
Elizabeth Potter

FEMINISM AND PHILOSOPHY OF SCIENCE

An Introduction

Elizabeth Potter

LONDON AND NEW YORK

First published 2006
by Routledge
2 Park Square, Milton Park, Abingdon, Oxon, OX14 4RN

Simultaneously published in the USA and Canada
by Routledge
270 Madison Avenue, New York, NY 10016

Routledge is an imprint of the Taylor & Francis Group, an informa business

Transferred to Digital Printing 2006

Typeset in Joanna by Taylor & Francis Books

British Library Cataloguing in Publication Data
A catalogue record for this book is available from the British Library

Library of Congress Cataloging in Publication Data
Potter, Elizabeth.
  Feminism and philosophy of science / Elizabeth Potter.
    p. cm. -- (Understanding feminist philosophy)
  Includes bibliographical references and index.
  1. Feminism and science. 2. Science--Philosophy. 3. Feminist theory.
4. Empiricism. I. Title. II. Series.

  Q130.P68 2006
  500'.82--dc22

                                              2005030911

            ISBN10: 0-415-26652-1 (hbk)
            ISBN10: 0-415-26653-X (pbk)
            ISBN10: 0-203-64666-5 (ebk)
         ISBN13: 978-0-415-26652-9 (hbk)
         ISBN13: 978-0-415-26653-6 (pbk)
         ISBN13: 978-0-203-64666-3 (ebk)

CONTENTS

ACKNOWLEDGMENTS

I wish to thank Indiana University Press for permission to use extended quotations from Chapter 4 of Bonnie Spanier's 1995 *Im/partial Science: Gender Ideology in Molecular Biology*, and Springer for permission to use extended quotations from Lynn Hankinson Nelson's "Feminist naturalized philosophy of science," *Synthese* 104(3) (1995): 399–421.

INTRODUCTION

The 1980s saw renewed attention by feminist scientists and science scholars to highly publicized findings of the 1970s and early 1980s that boys are better than girls at mathematics and other areas necessary for educational and economic success. Many scientists at that time interpreted their findings as explaining the prevalence of men in higher-paying and higher-status jobs.

For example, in 1973, to explain the fact that "there were more institutionalized mentally retarded males than females," Dr. Robert Lehrke hypothesized that "a number of genes related to intellectual ability reside on the X-chromosome and that, because of the peculiarities of chromosomal inheritance, X-linkage means that males will exhibit greater variability in intelligence." There are many possible explanations of this fact, but Lehrke resurrected an assumption that had been a popular explanation of greater (white) male educational and economic success from 1903 through the 1930s, viz. intelligence in men is more variable than in women. This meant that, although men and women have the same average intelligence, the number of men with high and with low intelligence is greater than the number of women with high and low intelligence. (Relatively fewer women have high and low intelligence.) If the hypothesis were true, Lehrke argued that it would explain why more men than women were institutionalized for mental retardation, viz. there would be a number of men with very low intelligence and few women with intelligence that low. The big news, however, was that the hypothesis also supposedly explained why men were more educationally and financially successful than women.[1]

In 1974, psychologists Eleanor Maccoby and Carol Jacklin published a survey of the literature pertaining to alleged cognitive sex differences such as verbal ability, mathematical ability, competitiveness, aggressivity, dominance, nurturance, and many more (Maccoby and Jacklin 1974). Of all the alleged differences, Maccoby and Jacklin noted that there were clear findings in only four: aggressivity and verbal, mathematical, and visual-spatial abilities.

Using the best methods of statistical analysis at that time, Maccoby and Jacklin concluded that, although the majority of studies showed no sex difference in verbal ability, nine studies did show a difference beginning at adolescence: eight showed that females are better at verbal reasoning. Only one showed that males are better. Fausto-Sterling notes that scientists can legitimately differ over whether to conclude, as Maccoby and Jacklin did, that females have greater verbal ability than males or to conclude that there is no discernible difference. She also points out that using meta-analysis (a more recent approach) on the same data, we find almost no difference. One scientist calculated a variance of "only about 1%" (Fausto-Sterling 1985: 29–30 (citing Hyde 1981)).

Turning to visual-spatial ability, Fausto-Sterling points out that more than half the studies show no sex-difference. In those that do find a difference, the difference favors males, but is very small, i.e. "no more than 5% of the variance" is due to sex. "The other 95 percent of the variation is due to individual differences that have nothing to do with being male or female" (Fausto-Sterling 1985: 32–3, citing Hyde 1981). Many of the tests of spatial skills are methodologically suspect, e.g. visual rod and frame tests (whose findings are contradicted by tactile rod and frame tests), but Fausto-Sterling notes that other tests are used as well: embedded figures and relevant parts of the Wechsler Intelligence Scales, especially the block design and mental rotations tests, angle-matching tasks and maze performance. Hyde calculated that 7.35 percent of males and 3.22 percent of females score in the ninety-fifth percentile, and assuming that the tests reveal abilities necessary for engineers and mathematicians, we would expect to find twice the number of males as females among engineers and mathematicians. (In fact, the number of US women in engineering is not nearly 50 percent; it is now closer to 3 percent!)

Fausto-Sterling remarks on the number of theories designed to explain the findings of sex-differences in verbal and spatial abilities that came and went between the 1960s and the 1980s. Of these, researchers are still actively investigating two: that hemispheric lateralization in male and female brains is different and that mathematical and spatial abilities are X-linked, i.e. males are more likely to inherit the X-chromosome carrying genetic information for mathematical ability and the information for spatial ability. Females are less likely to inherit the two X-chromosomes carrying genetic information for mathematical ability and the information for spatial ability. (See Fausto-Sterling 1985 for explanation and criticism of these theories as of 1985 and Fausto-Sterling 2000, especially chapters 5 and 9, for further discussion and criticism of more recent work on hemispheric lateralization and genetic influences on gender differences.)

When these theories are announced, they usually receive broad media attention (but, as Fausto-Sterling notes, when they are refuted or languish, there is no such attention). The fact that these sorts of theory come and go with such regularity begs for attention from feminists, not only to the specific hypotheses and claims put forward by scientists and trumpeted in the media, but to the nature of the sciences themselves and our understanding of them. Are social and moral values at work when these scientists choose their questions and methods, and interpret their findings? Are scientists influenced in their work by the cultural assumptions of their place and time? Supposing values are in play, do these values necessarily make the work bad science? Does it make the work relativistic? Good scientists are objective and their work is objective, but what is objectivity? Does it have one clear meaning accepted by all scientists and all philosophers of science? These are the philosophical issues raised by cases such as those described above.

Feminist philosophy in general represents the intersection of two inquiries: philosophy's inquiry into important domains of human life such as the moral, political, social, epistemological, and religious domains, and the inquiry of feminisms into the causes of and solutions to the systematic subordination of women to men. Feminist philosophy of science focuses both of these important inquiries upon the sciences. Using new and traditional approaches to philosophy of science, this sub-discipline has begun with inquires into

1   the social causes of women's subordination and the putative natural inferiority of women as the sciences understand them,
2   the contribution of the sciences themselves to women's subordination,
3   the contribution of philosophies of science themselves to women's subordination, and
4   philosophies of science that allow 1–3 to constitute proper subject matter for philosophy of science.

Philosophies of science, of course, are never static. They undergo continuing criticism and change as philosophers seek the best ways to understand the relationship between humans and the world and the production of knowledge about the world. Feminist philosophers engage in this process of criticism and change *qua* philosophers of science and *qua* feminist philosophers of science. In pursuit of 4, philosophies of science that allow 1, 2, and 3 to constitute proper subject matter for philosophy of science, feminist philosophers, as we shall see in the following chapters, have not begun *de novo*, from scratch, but have taken up existing approaches and modified them. Those familiar with both Anglo-American and continental epistemology and philosophy of science will recognize the underpinnings of each approach discussed in this book.

Seeking philosophies of science that facilitate inquiry into 1–3, feminists have focused strongly on the relationship between science and values. This focus has in large measure determined which issues within philosophy of science feminists have had to work on first. These include issues surrounding values and objectivity, judgmental relativism, pluralism, underdetermination, the agent of scientific knowledge – with particular emphasis on epistemological individualism vs. scientific knowledge as a social process and product, and the "situated" nature of epistemic agents. We will find these issues addressed in every feminist philosophy of science, albeit with very different responses to them.

A review of the various responses reveals that there are no bright lines between feminist approaches to these issues. The division of feminist epistemologies into three categories, empiricism/standpoint theories/postmodernism, introduced by Sandra Harding, has been very important for understanding this body of work. According to feminist spontaneous empiricism, existing methods and norms are sufficient to provide empirically and theoretically adequate research results. Failure to follow standard methods and norms leads to sexist and androcentric results (Harding 1993: 51). According to feminist standpoint theories, knowledge claims do not break free of their "original ties to local, historical interests, values, and agendas," yet they give a "faithful account of the 'real' world." Existing methods and norms are not sufficient to provide adequate research results and must be strengthened (Harding 1993: 50 citing Haraway 1991a). Finally, feminist postmodernism rejects claims to universal knowledge about "the existence, nature and powers of reason, progress, science, language and the 'subject/self.'" Instead, feminist claims are more plausible and less distorting "insofar as the are grounded in a solidarity between . . . modern fractured identities [e.g. black-feminist, socialist-feminist, etc.] and between the politics they create" (Harding 1986: 28 citing Flax 1986).

I believe that we are well served now by a new conceptual framework comparing approaches to central issues. For example, as we shall see in Chapter 5, the central insight of standpoint theory, that epistemic advantages accrue to knowledge claims arising within developed standpoints, fits well within feminist empiricist epistemologies of science. I suggest a nuanced differentiation of epistemic stances toward justification, epistemic agency, values and objectivity, underdetermination, pluralism, judgmental relativism, and other themes. Within this framework, we can still divide feminist epistemologies according to the tricotomy, but we can also divide them in different ways, for example, by how naturalizing and how normative they are. The chapters of this book are set up so that we compare feminist philosophers of science by their responses to the common issues facing them.

Feminist epistemologists and philosophers of science have engaged in lively debates over which feminist philosophies of science are least problematic and which best facilitate approaches to the social causes of women's subordination and the putative natural inferiority of women as the sciences understand them, the contribution of the sciences themselves to women's subordination and to the contribution of philosophies of science themselves to women's subordination, as well as which are the best responses to the philosophical issues concerning values and objectivity, judgmental relativism, pluralism, underdetermination, and the agent of scientific knowledge. On the whole, feminist philosophies of science have been to a greater or lesser extent naturalized philosophies of science. As Lynn Hankinson Nelson characterizes naturalized epistemologies, they should

1   be commensurable with the actual history and contemporary practice of science, i.e. subject to the same criteria as the sciences,
2   be grounded in sciences relevant to theories of theorizing, e.g. empirical psychology, social psychology, cognitive science, evolutionary biology, and/or sociology, and
3   have consistent methodological principles for explaining consensus and dissent, progressive and less than progressive episodes in science;

i.e. explain consensus and dissent, progressive and less than progressive episodes, in the same terms. (See Chapter 1 for discussion of these characteristics.)

Naturalized philosophies of science, then, aim to be empirically adequate; that is, like scientific theories, they must conform to a rich body of facts about the sciences or have broad scope, i.e. apply to a range of cases in the history of science (including current science). Beyond this, feminist (like mainstream) naturalized philosophies of science are more or less naturalizing. In mainstream philosophy of science, for example, Quine and others advocate that epistemology and philosophy of science be a subset of some science, especially of empirical psychology and/or evolutionary biology. Yet other philosophers retain a stronger normative role for epistemology and philosophy of science. As they see it, this role is to distinguish good science from pseudo-science, rational from irrational science, and to distinguish better and worse candidates for scientific knowledge. I have arranged the chapters of this book in a way that puts feminist philosophies of science on a spectrum from more to less naturalized and less to more normative.

Feminist philosophers have found naturalized epistemologies and philosophies of science promising because they offer ways to understand important aspects of how scientific knowledge is actually produced. Because we use knowledge produced by the sciences every day to determine our commercial, military, health, educational, and other social arrangements and to justify our gender arrangements in these areas, feminists have been concerned to understand the sciences as they are practiced, not as non-naturalizing philosophers have rationally reconstructed them. To rationally reconstruct an account of scientific work is to use a logic of science or schema of science as the principle of selection for which facts to include in the account. Such accounts do not intend to be full or very adequate to all the facts. (See Chapters 1 and 6.) In part, feminist philosophers of science need empirically adequate accounts because, as feminists, they seek to uncover the sources of women's subordination. They also criticize these sources and seek changes that will end the subordination of women; therefore, feminist philosophies of science have normative dimensions. (These may be in addition to normative philosophical intuitions feminist philosophers share with other, non-feminist philosophers of science.)

Feminist beliefs, e.g. that most women in the world suffer unjust systematic discrimination (which is not incompatible with the successes women enjoy), and non-feminist beliefs, e.g. that they do not suffer unjust discrimination (because, for example, any discrimination against them is justified, e.g. by their biology, in the same way that short people are justly rejected as members of professional basketball teams), are both evaluative beliefs. Such evaluative beliefs are treated in philosophy of science as moral and social values. In the same way, women are understood to have interests in, among many other things, the improvement of their lives by the elimination of unjust discrimination against them, by an increase in knowledge about their biology, by improvements in their access to health care, education, well-paid jobs, child care, etc. In the US, the sciences are treated as the final arbiters of knowledge about all aspects of women as biological, psychological, and social beings because scientists are understood to be experts in these areas. This means that scientists are called upon to help society determine social policy concerning women; thus, we can see why feminist science scholars need to know whether and how moral and social values influence the knowledge produced by the sciences, especially whether and how they do so in *good scientific work*.

It is not surprising, then, to find that the role of values and interests in scientific inquiry is of central concern to feminist epistemology and philosophy of science. Proponents of the view that social values and interests have no place in the "content" of good scientific work (for example, in testing hypotheses) have several arguments, some of which will be discussed in the following chapters. Their basic concern is that when social values and interests, understood as biases, enter the laboratory, enter scientific research itself, the result is bad science. Feminists agree that this is sometimes the case; the question is whether this is always or even usually the result. In the absence of an a priori argument that (roughly) social values always cause bad science, no one is well served by philosophies of science that make it impossible to find out (see Potter 1995). All feminist philosophies of science aim to make it possible to determine whether and how social values and interests enter the "content" of good scientific work – also referred to as "the context of justification."

The context of justification is distinguished from the context of discovery and from the social uses of the results of scientific work. The context of discovery includes the actual circumstances under which a hypothesis or model is conceived, and, of course, personal, commercial,

and social values and interests play a part here. But the context of justification is supposed to include work confirming or disconfirming a
hypothesis or model, work carried out in accordance with the best relevant methods. This is the location of disagreement between feminist (and
other) philosophers of science who find that social values and interests can
influence scientific work and the work still be good work, and philosophers of science who do not believe that this is so or that it could be so.
Finally, everyone agrees that many interests and values determine how the
results of scientific research are used. The distinction between the context
of discovery and the context of justification is discussed in Chapter 1 and a
brief history of the distinction is offered in Chapter 6.

Advocates of the belief that "science is value-free" frame this belief by
arguing that science is not *objective* unless the context of justification is free
of moral and social values. In response, feminist philosophers of science
offer analyses of objectivity and accounts of objectivity that aim to be
empirically adequate to cases of good scientific work and compatible with
feminist values.

Elisabeth Lloyd argues that for the past few decades, contemporary
analytic philosophers have recognized that the meaning of "objectivity"
and of "objective" is not "transparent, simple, stable and clear."
Philosophers have, therefore, proposed many and various accounts and
definitions of it and in her analysis of these proposals, Lloyd identifies four
distinct meanings of "objective:"

Sometimes:

1   *Objective* means *detached*, disinterested, unbiased, impersonal, invested in
    no particular point of view (or *not having* a point of view);
2   *Objective* means *public*, publicly available, observable, or accessible (at
    least in principle);
3   *Objective* means *existing independently* or separately from us;
4   *Objective* means really existing, Really Real, the way things really are
    (Lloyd 1995: 353).

"Objective" in each of these senses is predicated of different entities:

1   is predicated of knowers; knowers are detached, disinterested, etc.
2   and 3 denote a relationship between reality and knowers; reality is
    observable, publicly available, etc. and reality exists independently of
    knowers; and

4   denotes a status of what is regardless of any relationship it has to
     knowers. Most philosophers also hold that properties and entities that
     are dependent on knowers, e.g. social institutions, Locke's secondary
     qualities (see below), or toothaches, have some sort of status as real,
     but do not have the same status as whatever exists separately and inde-
     pendently of knowers and/or are not Really Real.

In one important philosophical picture, all four definitions are at work:
"objective" characterizes a relationship between knowers and reality-as-
independently-existing, and, methodologically, the knower must be
detached, because investment in a particular belief or attachment to a point
of view – usually understood as bias – "could impede the free acquisition
of knowledge and correct representation of (independent) reality" (Lloyd
1995: 354). Biases interfere with our access to reality. This picture is
behind what Lloyd dubs "the ontological tyranny" (though the picture
itself can be understood in ways that do not necessarily embrace the ontological tyranny).

The ontological tyranny begins with "the strong claim that 'objective'
reality – the reality converged upon through the application of objective
methods – equals all of the Really Real." Moreover, all of the Really Real can
be known because it is publicly accessible to those of us who use these
objective methods and who are properly detached or disinterested.

There are several assumptions here:

a) the ontological assumption that the Really Real is completely indepen-
    dent of us; thus,
b) objective knowledge of this Reality requires an "objective method"
    characterized by detachment, because
c) attachment or point of view might interfere with our independence
    from the reality we wish to know, and
d) this reality is publicly accessible, if it is accessible at all (Lloyd 1995:
    354–6).

Lloyd shows us that the ontological tyranny originated in the seventeenth
century as scientists (e.g. Robert Boyle) and philosophers (e.g. John Locke)
sought to defend current physics, and in the eighteenth century it became
a standard for all knowledge. Lloyd dubs this standard of knowledge the
"philosophical folk view of objective knowledge." It arose from views

about primary and secondary qualities: primary qualities are the real material entities; they cannot be separated from bodies and they are absolute and constant; they are what are known and can be represented wholly mathematically. Secondary qualities, on the other hand, arise from, are mere effects upon our senses of primary qualities; they are fluctuating, confused, relative, and untrustworthy. They have a different ontological status than primary qualities have because their existence depends on some sensing being.

The distinction between primary and secondary qualities was made by particular historical individuals and communities with particular projects involving physics. Each of these scientists and philosophers had prior ontological commitments that justified their adoption of specific methods for investigating nature. Lloyd notes that they were explicit about their commitments; for example, Galileo stated:

> Philosophy is written in that great book which ever lies before our eyes – I mean the universe – but we cannot understand it if we do not first learn the language and grasp the symbols, in which it is written. This book is written in the mathematical language, and the symbols are triangles, circles, and other geometrical figures, without whose help it is impossible to comprehend a single word of it, without which one wanders in vain through a dark labyrinth.

Here Galileo says that the reality of the physical universe is geometrical, from which it follows that the Really Real must be unchanging and constant, and therefore subject to mathematical representation. Moreover, his commitment to mathematical science was based on a religious view, as was Descartes's. Galileo said, "As to the truth, of which mathematical demonstrations give us the knowledge, it is the same which the Divine Wisdom knoweth." God made the world an immutable mathematical system; therefore, scientific knowledge can be certain when humans use the mathematical method (Burtt 1932: 75 and 82 quoted in Lloyd 1995: 358).

Galileo and Descartes were already committed to "an ontology that was fundamentally religious," then argued for the distinction between primary and secondary qualities. Thus, the ontological tyranny originated from a philosophical view, not from scientific discoveries; moreover, it is a philosophical view resting on prior, unargued for, ontological commitments – indeed, on religious commitments.

Components of the ontological tyranny appear in philosophically popular updated forms; Lloyd dubs one of these "Type/Law Convergent Realism." This view holds that *real* objectivity will result in a convergence on One True Description of reality. Thus, *real* knowledge "carves Nature at its joints." This epistemological criterion for knowledge presupposes the metaphysical view that "Nature *has* joints, i.e., 'natural' objects and/or events, and kinds, and laws, which could serve (ideally) to guide inquirers" to discover them. "Type/Law Convergent Realism" rejects pluralism in all its many forms. We will discuss several versions in the following chapters, but, roughly, epistemic pluralism holds that Reality allows itself to be known through many categories, non-overlapping ones as well as overlapping, conflicting ones, and the categories we use reflect our interests and values. Under Type/Law Convergent Realism, it is not enough that Reality itself *allows* certain sets of categories of objects, events, kinds, or laws; these sets must conform to nature's own categories, individuals, and laws, otherwise we do not have true descriptions of reality.

Philosophies rejecting the ontological tyranny and Type/Law Convergent Realism thus abandon the view that knowledge of reality requires objectivity understood as "'neutral,' 'non-ideological,' and 'distanced from any personal interests or idiosyncrasies.'" Moreover, we find no consensus on the best alternative understanding of objectivity.

Two things follow from the absence of consensus about how best to understand objectivity when we have rejected the ontological tyranny.

First, we might characterize as holding a "double-standard" or as engaging in "bait and switch" those philosophers who themselves problematize objectivity and offer or seek redefinitions of it, yet demand that feminist philosophers recognize a view of it based upon the ontological tyranny or upon updated versions of the ontological tyranny, demand that feminists *not* problematize it, and *not* offer alternative understandings of it (Lloyd 1995: 375).

Second, Lloyd argues that once philosophers abandon Type/Law Convergent Realism, they must account for the socio-cultural issues that immediately arise. In addition to "resistances by reality, values and interests, broadly speaking, are necessarily involved in the development of knowledge and concept-formation." Many of the interests that direct us to certain phenomena arise because we are an animal of a particular size, with particular brains and senses, having certain needs, living in the communities we do with the many aims that we have. But Lloyd correctly points out that anthropologists are virtually unanimous in holding that "sex and gender roles lay the foundations of every human society's other social practices – including communication, lines of authority, distribution of physical, emotional, and intellectual goods, and the very general social structures of who decides what." Therefore, any epistemology and philosophy of science that includes social interests and values as integral to the acquisition of knowledge should include sex and gender-related values and interests. Without an a priori argument to support the exclusion of sex and gender, these epistemologies and philosophies of science are inconsistent. Lloyd discusses several philosophers who either ignore sex and gender while including less fundamental social practices that are constructed on the "bedrock social roles of male and female, masculine and feminine" or who give arguments for omitting some social practices from their accounts while dismissing sex and gender out of hand (e.g. Searle 1984 and 1992) (Lloyd 1995: 373 and 367–8).

Each of the feminist philosophies of science treated in this book offers a viable account of objectivity. These accounts are all social accounts of objectivity set in social accounts of scientific knowledge.

Social accounts of knowledge can best be read as denying epistemic individualism – understood by feminist philosophers as a general picture according to which autonomous individuals can produce knowledge. Turning to science, epistemic individualism is the view that individual scientists can and do autonomously produce scientific knowledge. Paradigm cases of the general picture are found in solipsism, for example as the rationalist Descartes presents it in his Meditations. But empiricists have also presented individualist accounts of how knowledge is produced; Locke is a classic example (see Chapter 1, Section 7). Most feminist epistemologists and philosophers of science hold, instead, that knowledge is produced and maintained by groups or by communities of people. We can begin to appreciate what it means to say that scientific knowledge is socially produced by attending to the necessity for peer review and the replicability of results. Hypotheses and models thought up by an individual scientist alone in her laboratory are not deemed facts that we know until they have been reviewed, accepted, and taken up by other scientists in her field.

Part of what makes her model or hypothesis acceptable to the relevant scientific community is that it is supported or confirmed by evidence arrived at using the community's standards and methods. (If she uses a new bit of methodology, it, too, must be argued for and accepted by her peers.) These standards and methods, in turn, are produced by scientific communities as they work to determine the best ways to find out what they want to know.

This, of course, is not to deny that individuals know things. But while there is consensus among feminist epistemologists and philosophers of science on the thesis that individuals are not the primary agents of knowledge in the sense that they can produce knowledge in isolation, there is no consensus on what it means to say that epistemic communities are the primary agents of knowledge. On this issue, we can fruitfully contrast the philosophies of science offered by Lynn Hankinson Nelson and Helen Longino.

There is, however, consensus among feminist philosophers of science that knowledge is "situated." Because knowers must learn and produce knowledge in epistemic communities, they learn and hold many (though not all) of the tacit as well as explicit beliefs of their communities. Knowers also live in social communities and share many of the values and interests of these communities. And it is precisely such social values and interests that traditional accounts of objectivity aim to rule out of good knowledge production. Philosophers of science who assume that bad science results from the entry of such values and interests into the context of justification argue that scientists must be objective in the sense distinguished by Lloyd as detached and disinterested; qua scientists, they and their work must be "value-free." Feminist philosophers of science, however, turn to case studies and sometimes do their own case studies showing that good science – by philosophically recognized characteristics, e.g. empirically successful, congruent with existing theory, having broad scope, simple or elegant, etc. – can be carried out in ways that assume social values and interests. In these cases, work in the context of justification is influenced by social values and interests. Feminists are particularly interested in scientific work resting on social assumptions about gender as well as race, class, sexuality, and other socio-political categories. Feminists have recognized that an individual scientist's values and interests stand a strong chance of being uncovered by his peers, but this is most likely to happen when his values differ from the values of the peers reviewing or participating in his area of research. Gender (and other such) values and interests tend to be widely held within a scientific community and can easily go unnoticed. Feminist philosophers of science propose, therefore, a variety of remedies.

The remedies are congruent with the particular complex of arguments put forward by each philosopher, as we see in this book. The complex includes an approach to (1) the balance between naturalizing and normative considerations, which is closely related to (2) the relationship between science and values, (3) the nature of objectivity and how to achieve it, (4) pluralism, (5) underdetermination, (6) how to characterize the agents of knowledge, and (7) the problem of epistemological relativism.

Lynn Hankinson Nelson argues that there is no bright line between scientific theories, philosophical theories, and common-sense theories. Thus, her philosophy of science takes a form of holism. It also meets the three criteria of naturalized philosophy of science set out above. This means that a good philosophy of science, like a good natural or social science, is distinguished by a balance among the norms of empirical success, predictive success, and explanatory power since these are the norms distinguishing good and bad science. It follows that epistemology in general describes and explains how knowledge is acquired and epistemology of science describes and explains how scientific knowledge is acquired. Neither justifies knowledge in general or our particular knowledge claims. A naturalized philosophy of science, therefore, gives an empirically adequate description of the production of scientific knowledge and asks whether the social processes currently characterizing scientific practices, recruitment, education, etc. are likely to produce the best theories and, if not, suggests changes to insure that they do. Feminist naturalized philosophy of science, then, attends to processes, including especially methodological processes, reflecting gender considerations. And since gender is never in fact separated from race, class, and other sociopolitical categories, feminist philosophy of science attends to these, also. Nelson's work attends to these considerations.

The hypotheses, models, and theories put forward in the sciences must be tested against the relevant evidence. And since Nelson, with the majority of philosophers of science today, does not recognize a sharp distinction between theory and observation, she argues that the evidence for a hypothesis includes the observational consequences of the hypothesis and a large set of theories within which the hypothesis is embedded, including common-sense theories. The hypothesis is related to many, but not all, of our current theories. These are all part of the evidence. Thus Nelson has a very broad notion of evidence, and offers a methodological principle for a naturalized feminist account of evidence: evidence is constituted by observations and theories, themselves supported by evidence and by other theories. Observations themselves depend on theories. We will explore her account of evidence in Chapter 1, Section 1.

The set of theories constituting evidence for a hypothesis or theory, Nelson argues (see Section 2), includes social values. Quine referred to these as "non-cognitive," suggesting that they are not corrigible, not subject to correction the way statements of fact are. He held to a sharp distinction between facts and values, and argued that "non-cognitive" values cannot enter the context of justification in good scientific work. Nelson suggested in 1990 that we reconsider the assumption that values, including political beliefs and theories, are not subject to empirical control and that there is no way to judge among them.

Moreover, she points to feminist and other case studies of good scientific work (discussed in Section 5) to show that socio-political values influence scientific justification. (Chapter 3 explores Elizabeth Anderson's model making clear the role of such values in science and the influence of science upon such values.) Nelson concludes from these case studies that socio-political values do not necessarily make science bad; lack of empirical success is what makes it poor science. We will take a look at one of the cases she sets out describing recent work in neuroendocrinology. The case concerns work on the hypothesis that hemispheric lateralization is sex-differentiated, and, like her other cases, shows us that evidence is holistic and that socio-political assumptions function as part of the evidence.

Feminist and other case studies also reveal that scientific knowledge is socially produced. In opposition to epistemic individualism, Nelson argues that knowledge is produced and maintained by communities − where communities are collections of knowing individuals, each depending upon the others in many ways. Individuals put forward candidates for knowledge and individuals do know things, but not autonomously, in epistemic isolation. Even an individual's beliefs depend upon language and theories learned from communities and must meet community standards of reasonableness. We explore these arguments in Sections 6–8.

Finally, work in the history and sociology of science makes clear that standards and methods in the sciences change over time with new discoveries, new theories and models, as interests shift, and for many other reasons. Standards and methods differ from domain to domain and, within the same domain, can vary locally, e.g. from laboratory to laboratory. Nelson offers many arguments, some of which we set out in Section 9, to show that this historical relativism does not entail pernicious epistemological relativism.

Alison Wylie's Consilience Model of Confirmation makes clear a version of holism compatible with Nelson's, and shows us how hypotheses and principles from different domains bear on the interpretation of data so that it functions as evidence for a hypothesis, on hypotheses taken as background assumptions supporting an argument from the evidence to the hypothesis under test, on the interpretation of the hypothesis, and, ultimately, on how hypotheses and principles from different domains bear on the theory giving rise to, and in turn supported by, the hypothesis being tested.

The Consilience Model of Confirmation is arrived at through Wylie's naturalized approach to the philosophy of science. Well-grounded in archaeology and working in the philosophy of archaeology, Wylie developed her model through analysis of many cases in this unique field of science. Her intense study has allowed her to see how archaeologists in fact judge the relative credibility of claims that something is evidence and of explanatory claims made on the basis of this evidence. We set out her model in Chapter 2, Section 4, and use some of Wylie's cases to illustrate it. (With Wylie's approval, I have formalized her model and given it this name.)

For many reasons, people living in the present want to know how people lived and thought in the past; this desire lies behind those domains of social science attending to the past, including anthropology and archaeology. Although archaeology can be done to uncover many of the lifeways of people who lived in the recent past, it is, perhaps, best known for work on people who lived in the distant past, especially people who left no written record or left symbolic records themselves in need of explanation and interpretation.

Both anthropologists and archaeologists attempting to understand past peoples must begin with some working assumptions about people. These assumptions, of course, include the range of assumptions they, as members of existing cultures, make about themselves. By complex and sophisticated analogies, they begin with their own everyday concepts and with the analytic concepts they use to understand their everyday concepts. In

Sections 2 and 3, we will explore Wylie's widely used model of "ethnographic tacking," which depicts how anthropologists and archaeologists argue from "us" to "them" by tacking from their own analytic concepts to the material consequences of past practices constituting the archaeological record. Archaeologists argue, using a repertoire of models of ethnographic practice, for hypotheses and theories about people in the past, making use of methods and results from the many other social sciences and the natural sciences that bear on their data. Thus, archaeologists must use analytical theories drawn from familiar sources to make hypotheses about past cultures and lifeways, and to interpret archaeological data as evidence to test these hypotheses. The problem immediately facing them is a particularly salient example of problems arising from the "theory-ladenness" of observations and so of observational evidence. If the evidence used to support hypotheses and theories about the cultural past is laden with theories based on the cultural present, how can we get from "us" to "them" without begging the question in favor of "our" understanding of them? We might think of this problem as trying to avoid epistemic anachronism, i.e. reading the cultural present back onto the cultural past.

The Consilience Model of Confirmation shows us how archaeologists handle this problem. As we shall see, it is reasonable for archaeologists to conclude that an interpretive hypothesis is adequate when independently constituted lines of evidence converge upon it. Mitigated objectivity is achieved when the ladening theories – middle-range, linking principles – used to construe/interpret data as evidence meet two technical requirements: they must be *secure* and *independent*. Archaeologists use an enormous diversity of evidence from the natural as well as social sciences and different lines of evidence mutually constrain a hypothesis when they converge or fail to converge on a coherent account of a particular past context.

Wylie argues that the mitigated objectivity she finds in archaeological standards of reasoning avoids the extreme epistemological relativism that would arise were different archaeologists simply to read different current assumptions onto the archaeological record. Her Consilience Model of Confirmation makes clear that archaeological evidence must be secure and independent in her technical sense of these terms. Once the evidence is secure and independent, then archaeologists can triangulate, set up a system of mutual constraint among different lines of evidence bearing on a hypothesis and ultimately on a theory. This method does not provide certainty, but it is certainly not arbitrary. (Wylie's Consilence Model is very promising and might well elucidate the confirmation of hypotheses and theories in the natural sciences as well as other social sciences.)

The model allows us to see where gender considerations enter archaeological reasoning, to see when the considerations do meet standards of good practice in the field and when they do not, for example when current assumptions about women and gender roles are uncritically taken up and used to support hypotheses about the cultural past in a particular context. Such middle-range, linking hypotheses about past women and gender roles fail to be *independent* of present assumptions and often fail to be fully *secure* (Section 5). We will briefly explore three cases in which gender assumptions are important components of archaeological reasoning. All are examples of the Consilience Model, revealing more or less consilience. They provide examples of good reasoning, i.e. providing more confirmation by collateral theories and linking principles; weaker arguments for a hypothesis, i.e. showing less consilience; and, in the worst case, the argument is viciously circular, thereby failing the requirement that collateral theories be independent because the argument uses a current gender theory in its attempt to justify a hypothesis purporting to be about past gender arrangements, but tacitly derived from the current theory.

Feminist philosophers of science representing very different approaches have independently arrived at or agreed with several views including the mutual influence of social values and the production of scientific knowledge, the problematic nature of the "ontological tyranny" and its associated picture of objectivity, the situated nature of epistemic agency, and others. Despite the fact that naturalizing, feminist empiricism recognizes the situated nature of epistemic agency, the phrase "situated knowledges" is usually associated with Standpoint Theory and standpoint epistemologies of science. The title of Donna Haraway's celebrated essay, "Situated knowledges: the science question in feminism and the privilege of partial perspective," captures two central elements, one upon which there is general agreement and one which has been very contentious among feminist epistemologists. While there is agreement that the production of good scientific knowledge is influenced by the assumptions scientists bring to it from their professional domains and from their social lives, feminists have not agreed that some of these situations and standpoints are epistemically privileged over others (see Hekman 1997).

Wylie is the first naturalizing feminist empiricist to offer an explicitly standpoint epistemology of science, bringing together central elements of feminist standpoint theories with feminist empiricist philosophy of science. We treat this portion of her work in Chapter 5, Section 10 as part of our treatment of standpoint philosophies of science. We shall see that Wylie offers an empiricist account of objectivity and argues that objectivity is improved in some domains by particular standpoints.

Wylie agrees with Lloyd that objectivity is understood in many ways. When it is taken as a property of knowledge claims, Wylie argues that it marks a claim as one that maximizes some combination of epistemic virtues. Empirical adequacy – or one of its close relations such as truth – almost always appears on lists of epistemic virtues and Wylie plumps for a notion of empirical adequacy that includes empirical depth, "fidelity to a rich body of localized evidence" and/or empirical breadth, the ability of a claim "to extend to a range of domains or applications." As a naturalizing philosopher of science, Wylie points out that the particular combination of virtues which is maximized depends on the specific epistemic project in hand. Thus, like Nelson, she points out that we cannot know, until we examine a case of hypothesis confirmation, which of the epistemic virtues the objectivity of the hypothesis requires in addition to empirical adequacy. (Wylie mentions the epistemic virtues of internal coherence, consistency with well-established collateral bodies of knowledge, and explanatory power, among others.)

We will set out Wylie's distinction between social location and standpoint, and examine her argument that particular social locations and particular standpoints can provide those who inhabit and develop them with epistemic advantage as they attend to certain domains. For example, having a social location as a woman and/or having a feminist standpoint can, in domains in which gender assumptions are consequential, allow scientists to contribute to the objectivity of models, hypotheses, and theories by spotting assumptions about gender hidden from other scientists working in the domain, spotting failures to consider evidence relevant to assumptions and hypotheses about gender, suggesting more objective hypotheses, etc. Wylie gives us many examples of the epistemic advantage in archaeology that accrues from employing a "gender lens" (Chapter 2, Section 5). But *gender* location and/or having a *feminist* standpoint do not always enhance the objectivity of scientific hypotheses; other locations and/or standpoints might be better suited to do so in some cases. Wylie's naturalizing approach is made clear in this point. As she notes,

> The question of what standpoints make an epistemic difference
> and what difference they make cannot be settled in the abstract, in
> advance; it requires the second order application of our best
> research tools to the business of knowledge production itself. And
> this is necessarily a problem-specific and open-ended process.
>
> (Wylie 2004: 40)

Feminist philosophers of science, of course, do not argue that there is no
distinction between facts and values as we ordinarily talk and think about
them. Rather, with Nelson, these thinkers are skeptical of received views
about the distinction found in epistemologies and philosophies of science.
Thus, feminist epistemologies and philosophies of science can best be understood as urging that philosophers should break down not the distinction
between "non-epistemic" (e.g. social, moral, political) values and facts as
determined by science, but break down the view that facts, the result of scientific work, cannot influence values, and the view that "non-epistemic" values
should not influence the production of scientific knowledge. Sometimes
these opposing views (and other, related views) are abbreviated as "there is a
sharp distinction between facts and values" and "there is no sharp distinction
between facts and values." We find an excellent statement of and argument for
the feminist proposal in the work of Anderson.

In Chapter 3, we will see, first, how Anderson renders the arguments
made by some critics against the view that social values and interests often
constrain and are constrained by good scientific work, that the two often *do
and should* influence one another. Because several of the premises in the
critics' arguments are suppressed, Anderson does the work of constructing
full and valid arguments for them, focusing on the arguments of Susan
Haack (1993). Critics argue, especially against feminist versions of the view,
that feminists hold their values dogmatically. Indeed, everyone holds their
values dogmatically because values cannot be influenced by facts. Moreover,
when social values and interests influence scientific work, they supplant
reasoning from evidence and so interfere with the goal of science, finding
out the truth about nature and about humans and human life, or finding
theories with a "good fit," i.e. that fit the data reasonably well. Thus, when
social values or interests influence scientific work they make it bad science.

Behind these arguments, Anderson detects "an unexamined cynicism"
about value judgments, viz. that value judgments only express wishes or
desires and are "not subject to critical scrutiny or revision in light of argu-

ments and evidence." She points out that no serious contemporary moral theorist "accepts such a crude emotivist account of value judgments" (Anderson 1995b: 35). Moreover, she argues (Section 6) that even emotional states can be warranted or not depending on the facts, and so warranted value judgments depend on facts. As we shall see, she uses this insight to set up her Co-operative Model of Theory Confirmation and sets out an excellent example to illustrate it.

Anderson is specifically responding to critics who read feminist empiricist epistemologists and philosophers of science as saying that scientists may choose the politically preferable theory when the facts underdetermine theory choice. Underdetermination has been treated as a *problem* to be solved by epistemologies and philosophies of science. We will discuss it in Chapter 4, Section 9, but, briefly, the problem arises when philosophers recognize that there are no "raw data;" instead, data are always "cooked," that is, theory-laden. Thus, there are no theory-neutral, unequivocal observations available to determine which of two or more competing theories is best. The many lists of "cognitive values" or "epistemic virtues" put forward by philosophers of science – usually including coherence with existing theories, fruitfulness, simplicity, internal consistency, scope, and the like – are to be used in deciding between competing hypotheses when both are empirically adequate to the data.

There are many versions of the underdetermination thesis; for example, Quine's version basically says that for every hypothesis and theory there is in principle an empirically equivalent hypothesis or theory. (As we shall see, Longino has a very different version of underdetermination.) It follows from holism, i.e. the view that our beliefs form a web connecting them all together, and from the underdetermination of theory by data, that we can hold on to any belief so long as we are willing to redistribute truth values across our other beliefs as needed to maintain a coherent system. Thus, Mary Hesse refers to the principles we use in the redistribution of truth values as "coherence conditions" (1974). Here "cognitive values" or "epistemic virtues" are understood as coherence conditions.

Though Hesse limits her list to certain metaphysical assumptions as well as assumptions such as the goodness of symmetry and of certain analogies, models, and so on, I have argued that any belief can function as a coherence condition; thus, androcentric and sexist beliefs can and do function this way in scientific reasoning. They, too, are beliefs that scientists use as background assumptions in the light of which they decide how to interpret evidence and maintain a coherent theory and set of theories. I have argued that feminist scientists can use feminist assumptions in the same way and still produce empirically adequate, epistemically virtuous theories. But, as Anderson points out, some critics have taken this to mean, "When the evidence underdetermines competing hypotheses, scientists may choose whichever one is politically preferable!" This is very different from choosing one that functions as a coherence condition *and* maximizes empirical adequacy and other epistemic virtues *and* is politically preferable, for example enables politically preferable solutions to medical or other social problems (Potter 1988, 2001). Anderson gives us a model of theory confirmation that provides an account of how value judgments can properly figure in theory choice, which does not just come down to choosing a theory because it is politically preferable. (We set out the model in Section 2.) Her account defends the view, to be investigated in Section 7, that non-epistemic values play legitimate and illegitimate roles as background assumptions mediating inferences from data to theory.

An important part of Anderson's argument is a critique of many critics' claim that the goal of science is truth and so theory choice must be made on the basis of truth alone, or truth-conducive considerations alone. She argues that even *true* statements about a given phenomenon can constitute a distorted, biased representation of it. What constitutes an adequate, unbiased representation of the whole depends upon our values, interests, and aims, some of which have moral and political import. Significance and lack of bias are legitimate criteria of theory choice, from which it follows that non-epistemic, context values play a legitimate role in justifying theories. These arguments are explored in Sections 1, 4, and 7.

Anderson argues that clear definitions of *significance* and *impartiality* reveal the need for values in the justification of theories for, without them, research has no direction. Mere truth is not enough to direct inquiry, and at worst it allows biased but true representations; it can allow simply a random collection of facts; at best it provides partial but harmless theories. Her case study, briefly described in Section 8, provides an excellent example of the way in which explicitly feminist values can help produce science that is better by standard scientific criteria.

Longino recognizes herself as a hybrid philosopher, insisting upon both naturalizing and strongly normative elements in her epistemology and philosophy of science. We will make both these elements clear in Chapter 4.

To show that social values and interests can and often do operate in scientific work that is accepted and well-respected by other scientists, Longino argues that scientific knowledge is produced through practices carried out primarily by communities of scientists, that background assumptions are always at work in scientific reasoning from evidence to hypotheses, models, and theories, and that contextual values and interests – including assumptions about gender and gender roles – can and often do function as background assumptions.

In 1990 Longino argued that there is no unique or intrinsic evidential relation; in particular, she agrees with the majority of philosophers today that the attempts of logical empiricists and post-positivists to find a logical relation between propositions expressing hypotheses or theories and those expressing data – primarily construed as observational – failed. Thus, scientific hypotheses, models, theories, etc. are underdetermined by sensory perceptions in the sense that perceptions *alone* tell us nothing about hypotheses, models, and theories. Scientists determine which perceptions function as data relevant to a hypothesis and so as evidence for a hypothesis *on the basis of* some of the background assumptions they hold. Background assumptions also determine in what way the data support a hypothesis.

Longino argues that these background assumptions include, as we have seen, what some epistemologists refer to as "cognitive values," e.g. truth, accuracy, simplicity, predictability, and scope, which are generated from an understanding of the goals of science, and are the source of rules determining what constitutes acceptable scientific practice. Longino rejects the terms "cognitive" and "non-cognitive" because the latter begs the question whether social values can function in the way that a list of technical, "cognitive values" does in rational scientific decision-making. Instead, she distinguishes such cognitive or epistemic factors as *constitutive values*. The other category she dubs *contextual values*, and includes in it personal, social, and cultural values and interests, e.g. expressions of preference about what ought to be. Although traditional empiricist philosophies of science contend that contextual values undercut or ruin otherwise good scientific work, Longino distinguishes five ways in which they *can* influence good work and she describes several cases in which they *do*. We will explore two of these cases in Chapter 4, Section 3. Both cases reveal evaluative beliefs about gender functioning as background assumptions in good scientific work.

Over time and occasionally during the same period of time, different scientific communities arrive at different, sometimes conflicting, theories and models of the same (or what appear to be the same) phenomena. Desire to avoid the radical epistemological relativism suggested by this state of affairs provides one of many reasons philosophers of science search for general norms to determine the superiority of one theory or set of theories. (Often, the search arises from desire for the One True Description of reality). Longino agrees with these traditional philosophers that we should avoid radical relativism and she has worked to find norms that distinguish knowledge from opinion. To dub something (e.g. the findings of scientific research) "knowledge" implies that we ought to believe it. In a favorite example, suppose Maya's reading tea-leaves is as empirically successful, e.g. as predictive, as an accepted scientific hypothesis. Longino argues that the results reached in a research project warrant the status of knowledge because the research leading to the results adequately meet four norms, whereas Maya's tea-leaf reading does not. The norms constitute requirements for the effective criticism without which science and other knowledge-producing practices do not meet our shared understanding of knowledge. We treat these four requirements in Sections 10 and 12 and we shall see that they would, if met, insure that feminist perspectives, criticisms, and contributions are considered in the practices of the sciences. One of the requirements, Tempered Equality, clearly calls for a diverse population of scientists holding diverse perspectives, such as feminist and anti-racist perspectives, insured through recruitment into the educational tracks leading to careers in science.

Longino departs from traditional empiricism in rejecting epistemic individualism and in rejecting the ontological tyranny. She argues that scientific knowledge, like most knowledge, is social, guided by social norms and produced in discursive interactions, and she adopts a version of pluralism in place of monism, which takes the aim of science to be finding the One True Description of reality using only one general approach. In Section 6, we will set out her argument that the natural world is too complex to be captured through a single theoretical approach and pictured in one unified account. Different accounts, including those that conflict, should be understood by analogy with maps having different projections. And conflicts between theories should be resolved, if possible, by limiting the domain of application of at least one of the conflicting theories. Since this is not always possible, research in some domains might go on until humans die out.

Because what counts as evidence for a hypothesis and how strongly it supports the hypothesis are relative to the background beliefs of the local scientific community, the objectivity of scientific inquiry appears to be in jeopardy. Longino's response is to argue that the social nature of science provides such objectivity and, in Section 12, we will examine her account of objectivity as the result of three ways in which scientists criticize the work of their peers: evidential criticism, conceptual criticism, and criticism of background assumptions. Here we will see how the four norms of effective criticism contribute to the objectivity of the sciences.

Inasmuch as the four norms are intended to function along with the basic empiricist norm, empirical adequacy, to demarcate knowledge arrived at rationally and objectively, Longino proposes new definitions of knowledge, including a definition of (1) the content of knowledge, (2) knowledge-producing practices, and (3) the knower. First, having a justification is understood to be the basic mark that distinguishes knowledge from opinion. Of course, the justification cannot be whatever an individual takes to be the justification of her opinion, so Longino uses the phrase "epistemic acceptability" instead of "justification." She is proposing a new definition of epistemic acceptability demarcating (1) content that is the content of *knowledge*. We treat the technical components of her definition in Sections 13 and 14. Her view of epistemic acceptability leads her to a very useful new success category to replace both correspondence and coherence theories of truth as the one mark of epistemically acceptable beliefs or knowledge. Instead of truth, Longino proposes the category of *conformation*, of which truth is only one mode.

Second, Longino proposes a definition of (2) knowledge-producing practices as those practices leading to the production of epistemically acceptable content; and, third, a definition of knowledge understood to be (3) an attribute of a knower (Section 14). This definition corresponds to the familiar model stating that "S knows that p if S believes p, p is true, and p is justified." Despite their surface similarity, we shall see that Longino's definition is a radical departure from the traditional model.

In Chapter 5, we will set out two standpoint epistemologies of science. "Standpoint" is a technical term in feminist philosophy, not synonymous with "viewpoint" or "perspective," and it is used in a way that distinguishes it from a "social location." In any socio-political order, we can usefully differentiate many social locations, all determined by the way a society orders the relations among people. In the US, for example, people are related to one another through structures built around gender, race, economic status or class, national origin, religion, and around many other socio-political categories. Each specific sub-category (e.g. masculinity, femininity, transgender, and other interesting sub-categories of gender) and their many intersections (e.g. working class, gay, Catholic Chicanos) constitute a social location. The people who share a location might share a perspective, but they do not thereby share a standpoint. A standpoint is an achievement, the result of analysis by a group of people, often people who share a social location. Harding argues that, in the first instance, a standpoint arises when people occupying a subordinate social location analyze the conditions of their lives and engage in political struggle to change them. In Chapter 5, Section 1, we will explore the distinctions among standpoint, location, and perspective sketched here, and we shall see that Harding offers not a particular standpoint analysis, but a standpoint theory.

Harding's standpoint philosophy of science is firmly rooted in feminism as well as anti-racism and post-colonial thought, and has important roots in Marxism. In the 1970s, Marxism provided a well-developed alternative to both rationalist/empiricist (positivist) epistemologies and methodologies and to "interpretationist" oppositions to them, Harding tells us. She and other feminist scholars who began to produce what came to be known as "standpoint theories" rejected the common view that gender relations are the result of individual choices made by autonomous individuals. Instead, they sought a means to understand the ways in which relations between men and women are institutionalized and a means to understand why feminist and androcentric thinkers produced such radically different accounts of these relations. Many feminist thinkers found Marxian insights useful for understanding these phenomena.

Thus, in Section 2, we will see that Marxian feminists could use the notion of "ideology" to explain dominant accounts of relations of gender, race, and class, and to show how groups with different gender, class, and racial locations tend to produce different accounts of nature and social relations. With this in mind, we understand how standpoint theorists could generalize from the proletariat, the economically subordinated group under capitalism, to women as the subordinated group under patriarchy. Dominant ideologies explain class and gender relations in ways that legitimate economic, gender, and other socio-political hierarchies. Of particular importance is the Marxian insight that the material conditions of peoples' lives can actually shape their understanding of the social and natural world. The extent to which this is so and the ways in which it occurs are questions to which many thinkers have addressed and continue to address themselves.

The proletarian, as an agent of knowledge, is presented in Marxian epistemology as the equivalent of the modernist rational man, well exemplified by the scientist. As a scientist, a man is objective in the sense that he has no point of view. He represents the universal knower: anyone without a point of view and trained in the proper methods will come to know the same things. Any social differences among these epistemic agents are epistemically irrelevant. Feminist standpoint epistemologists, as we shall see in Section 3, argue that agents of knowledge are, on the contrary, "situated," i.e. even the content of their thought is shaped by their social locations. Thus, for feminist standpoint theory, in contrast to the Marxian and bourgeois, modernist views, knowing agents, including scientists, are local and heterogeneous.

Harding also argues (Section 4) that communities, not individuals, produce knowledge. And she points out two senses in which the community as the primary agent of knowledge is heterogeneous as opposed to homogeneous. First, different epistemic communities – of which scientific communities are a good example – can differ in many ways from one another, and although different epistemic communities might produce similar, compatible accounts of the same domain of the natural or social world, they can produce conflicting accounts. Second, all epistemic communities are, like scientific communities, internally heterogeneous inasmuch as they are made up of people who are significantly different from one another.

Two problems arise from these ways in which epistemic communities are heterogeneous. If scientists are not epistemically the same and are local in the sense that their thoughts are shaped, though not determined, by the material conditions of their lives, how can they be objective in the way that science demands? Too, if different scientific communities investigating the same domain produce conflicting theories accounting for the domain, how can scientists objectively decide among them? How can we avoid radical epistemological relativism in our accounts of scientific theory choice?

The sciences have developed methods to prevent biases from distorting the results of research. But Harding refers to the objectivity provided by such methods as "weak objectivity" and proposes, as we shall see in Section 5, not that those methods be replaced, but that scientists create ways to insure "strong objectivity." Existing methods try to insure that an individual scientist's interests, prejudices, and personal values do not bias the results of research, but, as feminist science scholars have argued, traditional standards of objectivity are too weak to identify beliefs, interests, and values widely shared by members of a research community. Harding argues that good science also requires these widely held assumptions to be revealed and examined, and this requires more democratic knowledge procedures, including especially procedures to ensure diverse scientific communities. In particular, when women within marginalized groups as well as women within dominant groups struggle against their oppression and achieve their own standpoints, these standpoints can contribute to the strong objectivity of scientific accounts. If the standpoints are used to critique dominant accounts of nature and of the social world, they can reveal hidden androcentric, Eurocentric, or class-based assumptions.

Harding offers many arguments to show that women have distinctive standpoints on the natural world. In Section 6, we set out some of these arguments and, in Section 7, we explore her discussion of the ways in which the standpoints of women can serve as resources for the sciences. Harding also proposes that science take up research projects that "start from women's lives." She argues that the experiences and lives of women (and of marginalized men), *as they understand them*, provide scientific problems to be explained and research agendas that differ from the problems that appear in dominant frameworks. When the perspectives of women and other marginalized groups help determine research projects and when the standpoints of women and the groups of which women are always a part contribute to the strong objectivity of science, science will be better in the technical sense that it will be less distorted and less false

To spell out her account of what it means for a theory or hypothesis to be "less false," Harding turns to the intriguing argument of N. Katherine Hayles (1993) that categories of epistemic assessment include true, not-true, less-false, and false. If "true" is reserved for hypotheses that, as Popper argued, pass all possible tests and "false" is reserved for hypotheses that fail important tests, then the best science can find in any domain is a theory that is less-false than all the others against which it has been tested. Having argued that different scientific communities investigating the same domain can produce conflicting theories accounting for the domain,

Harding's standpoint philosophy of science appears to be epistemologically relativist. But she avoids epistemological relativism by rejecting the incommensurability of conflicting theories through an appeal to Galison's (1996) notion of the sciences as having boundaries across which scientists can communicate using "pidgin languages." The ability to communicate in this way allows different scientific communities to argue rationally about which account is less-false. Of course, there is no guarantee that they will ever agree, but the problem here is not relativism. We investigate these arguments and distinctions in Section 8.

Finally, Harding rejects the ontological tyranny, arguing that there is not always one "less false" theory, and two or more theories may be incompatible yet deemed by scientists to have passed the relevant tests. In Section 9, we shall see that she expresses her pluralism as the possibility that in any domain there might be different *partial* representations, but notes that "many highly useful but *conflicting* representations can be consistent with 'how the world is,' although none can be uniquely congruent with it" (Harding 1997: 383; italics mine). Thus, Harding allows for a plurality of theories all of which can be empirically adequate and useful, some of which may be compatible, but some of which may contain knowledge claims that are "fundamentally incompatible." She cites approvingly the version of metaphysical pluralism described by John Dupré (1996) as "promiscuous realism."

Feminist empiricism and feminist standpoint theories have, to date, been deemed incompatible. But in Chapter 5, Section 10 we present a *feminist empiricist standpoint theory*, that of Wylie, discussed above. As we have seen, Wylie's theory is a standpoint theory recognizing the role of location in shaping knowledge and the useful role of standpoints and perspectives of women in determining research projects and critiquing science. However, Wylie's standpoint theory differs in significant ways from Harding's. As we shall see, Wylie argues that particular standpoints provide epistemic advantages in particular domains of science and she responds to the threat of epistemic relativism using a standard philosophical account of objectivity.

Finally, in Chapter 6 we face the criticism that feminist philosophies of science are not good philosophies of science because they are not "value-free" philosophies; as feminist, they adopt a standpoint; they are motivated by and employ feminist values and beliefs in their analyses and arguments.

In response, I will argue that philosophical skepticism about the possibility of value-free science is not new; it did not begin in the twentieth century. The skepticism was addressed, though not discussed, in 1938 by Hans Reichenbach as he formalized the set of distinctions still used today among the context of discovery and the context of justification; and between the division of labor between philosophers who do conceptual analyses of science and social scientists who do empirical studies. These are supplemented by the distinction between internal accounts of scientific rationality, to be put forward by philosophers and external accounts (of factors leading to irrational failures to achieve scientific knowledge) to be done by historians, psychologists, and sociologists of science. This apparatus functions to distinguish rational science from aspiring failures.

By 1953, the argument that science cannot be value-free was nicely put forward by Richard Rudner. We will briefly consider Carl Hempel's answer to Rudner. In the early 1980s, he addressed the claim "that science *presupposes* value judgment" and set out a list of epistemic values that mark those hypotheses and theories that it is rational to accept. (We could as well consider the arguments of Quine, Kuhn, or other philosophers of science who put forward similar lists of epistemic values.) Hempel hoped to find a theory of probability that would capture the essence of good reasoning from a given body of evidence to a hypothesis, and this theory of probability together with the list of epistemic values would constitute an elegant logic of science by which to distinguish rational from irrational scientific decisions.

Feminist (and mainstream) philosophical arguments that truth and probable truth are not the sole aims of science, that human values and interests determine the point of most scientific research, along with arguments that contextual values function in the same ways that constitutive values do and can help produce better science, thus fly in the face of a venerable philosophical tradition.

But the counter-claim that moral and social values can influence scientific methods and are not "biases" or are somehow not bad biases also flies in the face of a venerable tradition and basic training in science. Few *scientists*, indeed, would want to admit that they allowed moral values to influence the way they, for example, interpreted their data! I will make a distinction between instrumentally "good" scientific work and "morally good" work, and argue that when scientists think or speak of "good work" they are thinking and speaking of instrumentally good work that is morally value-neutral. Moral neutrality is a norm of good science.

I will then argue that it is also a metaphilosophical norm of good philosophy of science and suggest that we *ought not assume* that well-

30

respected philosophies of science (well-respected regardless of whether they are widely accepted or adopted) are in fact neutral among contextual values. We need to do the same kind of empirical research on philosophies of science that scholars have done on particular cases of scientific work to determine whether they are in fact free of contextual values, i.e. adopt no contextual values in their own background assumptions, etc. In the absence of an a priori argument showing that rational philosophy of science is necessarily neutral in these ways, and looking at just two of the many locations at which contextual values are very difficult to avoid in constructing a philosophy of science, we may conclude that feminist philosophy of science is virtuous, after all, since it wears its values and interests for all to see; it does not hide them in its closet.

1

NATURALIZED FEMINIST EMPIRICISM

Lynn Hankinson Nelson was the first to declare for a feminist natural-ized philosophy of science, one that meets three recognized criteria for naturalization. A naturalized philosophy of science should:

1   Be commensurable with the actual history and contemporary practice of science; we can take this to mean that a naturalized philosophy of science is subject to the same criteria the relevant sciences are subject to, for example empirical adequacy – understood as conforming to a rich body of evidence and/or as having scope or a range of applica-tions. And meeting criteria such as empirical adequacy is one reason naturalized philosophies of science require far less rational reconstruc-tion of history or contemporary practice than traditional accounts – although philosophers of science always have to select which facts to present when setting out cases. (Compare, for example, Conant 1970 and Potter 2001.)
2   Be grounded in sciences relevant to theories of theorizing, e.g. empir-ical psychology, social psychology, cognitive science, evolutionary biology, and/or sociology.
3   Have consistent methodological principles for explaining consensus and dissent, progressive and less than progressive episodes in science, i.e. adhere to a principle of symmetry, explaining consensus and dissent, progressive and less than progressive episodes, in the same terms (for example, if we give a social explanation for dissent, we should also explain consensus in social terms; thus, we do not argue that the consensus around successful theory T was reached solely because T is true, while the dissent was caused by the social values of the dissenters). And, she argues, philosophies of science attending to

the practices of feminist as well as non-feminist scientists and to results in feminist as well as mainstream science scholarship satisfy these three criteria better than those that do not (Nelson 1995).

1.1 A feminist empiricist account of evidence

Nelson develops Quine's empiricist philosophy of science in ways that make it feminist. (Of course, these developments produce an empiricism very different in many ways from Quine's.) We will not here describe Quine's philosophy of science; instead, we will begin with his conclusions that there are no foundations of knowledge and there are no pre-theoretic or untheorized observations, i.e. there is no sharp distinction between theory and observation (see Nelson 2000 and Nelson and Nelson 2003). Philosophers had counted on a sharp distinction based on the worry that if all observations are "theory-laden," then using observations as evidence for theories must beg the question in favor of those theories. This worry leads to foundationalism: for empiricists, there must be some sensory evidence, itself unconceptualized, that provides the ultimate justifying evidence for our knowledge claims. Thus, as empiricists, Quine and Nelson owe us an account of what evidence is that retains observations as the final arbiter of our theories without treating observations as foundations of knowledge. Nelson suggests the following as a methodological principle for a naturalized feminist account of evidence, FAE:

> The evidence supporting a specific theory, hypothesis, or research program is constituted by observation, itself largely structured as current theories would have it, and other theories that together constitute a current theory of nature, inclusive of those informed by social beliefs and values.
>
> (Nelson 1996: 100)

The acceptance of FAE marks Nelson's philosophy of science as empiricist; i.e. it says that evidence is constituted by observation *and* by theories that themselves are supported by evidence and by other theories. When Nelson says that observation is largely structured as current theories would have it, she means that observational experience is not explained by a philosophical account such as sense-data theory, but by *scientific* accounts such as those given in neurobiology, developmental biology, neuropsychology, and evolutionary biology. FAE also distinguishes Nelson's feminist empiricism, not only from many mainstream empiricisms, but also from other feminist empiricisms, most notably Longino's. This is because FAE entails a *very broad notion of evidence*, much broader than traditional empiricists have supposed. Evidence has been limited in traditional empiricist accounts to the deliverances of our senses or to empirical observations based upon our senses and often referred to as "data." And these data are held to be independent of any theories so they can support a hypothesis or theory without begging the question by depending on them.

As Nelson points out, the question whether theories and observations are independent of each other was settled many years ago. So the question now is "What are the implications for *evidence* of the demise of the theory/observation dichotomy?" How are hypotheses, theories, models, etc. tested and supported by sensory evidence? According to many post-positivists, a hypothesis faces the test of experience together with auxiliary assumptions (which include a large part of science). And according to semantic theorists, including Longino (1990), a hypothesis faces the test of experience in the form of data along with background assumptions that mediate the relationship between the data and the explanatory hypotheses, models, and theories. But according to Quine himself, a hypothesis faces the test of experience only together with *all* of science *and with our "common-sense" theories*. In this picture, there are no sharp boundaries between any of our theories; not between scientific theories and not between the theories of science and those of common sense. This is why Quine's philosophy is referred to as "holism." He describes all of our theories, scientific, philo-sophical, and common sense, as a "network." Nelson differs here from Quine, arguing that a hypothesis faces the test of experience together with a "modest chunk" of science and common-sense theories, not necessarily with all of science and all common-sense theories.

Thus, for Quine the unit of empirical significance and the test for a hypothesis or theory is a body of theory, in principle all current science; but for Nelson, the evidence for a hypothesis includes the observational consequences of the hypothesis and a larger chunk of the theories within which the hypothesis is embedded, i.e. the relationship of the hypothesis to many, *but not all*, current theories, metaphysical assumptions, methods, standards, and practices. These, of course, include "common-sense" theo-ries. Sometimes Nelson says "however much we can accommodate;" this means that the evidence against which a hypothesis or theory is tested includes as much current science and common sense as humans can actually apply when testing the hypothesis or theory. Thus, she leaves it an open question how large a chunk of theory and results is used in assessing a specific research program or theory; the "size of the chunk" varies from case to case (Nelson 1996: 101). But our current theories, metaphysical assumptions, methods, standards, and practices are all *part of the evidence*; they are not "merely background assumptions, auxiliary theories or 'disciplinary matrices.'" So her holism is a "modest holism" compared to Quine's view of evidence, but compared to Longino, Lloyd, and other semantic theorists, it is still holism with a very broad notion of evidence.

1.2 Facts and values

Nelson is much more rigorous than Quine was about the consequences of holism and of the coherence theory of evidence. Quine argued that there is no sharp boundary between common-sense theories and beliefs and scientific theories and beliefs with one important exception: he believed in a strong boundary between science and non-constitutive values. This is because he thought that moral, social, and other such values are not subject to empirical control. Nelson argues, to the contrary, that socio-political claims and non-constitutive (usually assumed to be non-scientific) values sometimes help constitute evidence in good science, i.e. science that scientists themselves say is good science. Thus, she formulates the coherence theory of evidence as FAE, according to which the theories that, along with our experiences, constitute evidence include values and socio-political theories. We must, she argued in 1990, "reconsider the assumption that political beliefs and theories, and values, are not subject to empirical control, that there is no way to judge between them" (Nelson 1990: 297). We return to this issue in our discussion of Anderson's model of the interaction between values and science.

Feminist and non-feminist science scholars have offered many cases, Nelson points out, showing that social processes, both internal and external to the "context of justification," social beliefs, and/or values, are sometimes part of good scientific work. What counts as a social or a political process or factor? Nelson points out that there are many candidates including "peer review mechanisms, funding mechanisms, 'negotiations' within science communities which lead to consensus, prestige hierarchies among sciences and specialties, the internal politics of disciplines and sciences, and features of larger social environments, including social relations of gender, race, and class" (Nelson 1995: 409; see also Solomon 1994 and 2001, Burian 1985 and 1993, Fuller 1988, Longino 1990 and 2002, Maffie 1991, Nelson 1990, Potter 1993 and 2001, and Stump 1992). Nelson refuses to give a list of factors because, as she says, "a naturalistic philosophy of science must allow the details of individual episodes to indicate which, if any, such factors were of import, in what ways, and to what degree." Traditional, non-naturalizing philosophies of science have responded to such case studies in different ways, by denying that such cases are good science, by denying that the case studies are correct, and/or by rationally reconstructing the cases to show why the social values or other factors are epistemologically irrelevant, i.e. not relevant to why they are cases of good science (Nelson 1995: 409 and 1996: 96–7).

1.3 Naturalized philosophy of science and its normative discontents

Feminist naturalized philosophy of science has no need to or interest in rationally reconstructing case studies of science in these ways. It is, therefore, more continuous with science and much more likely to be empirically successful.

One of the greatest concerns raised by naturalized philosophy of science is that, if philosophy is continuous with science, philosophy gives up its normative role, especially the role of providing criteria by which to distinguish good science from bad science. We should note, therefore, that Nelson's holistic view of evidence leaves norms intact: empirical success, explanatory power, and predictive success are still the basic norms distinguishing good and bad science. Thus, she introduces the example of the nineteenth-century science of craniometry, which looked for sex, race, and class differences among human beings; craniometry is now deemed to have been bad science. On Nelson's view, it was bad not because non-constitutive values were used as part of the evidential warrant for it, but because its observational consequences were not borne out. It lacked explanatory power and empirical adequacy (Nelson 1995: 406). We see, then, that on Nelson's view, when good values constitute part of the evidence for a hypothesis or theory, they do not make the science bad; nor do bad values make otherwise good science into bad science. The distinction between good and bad science is still based on traditional constitutive virtues including (though not limited to) empirical adequacy, explanatory power, and predictive power.

Epistemology and philosophy of science do, however, give up their traditional normative role because, Nelson argues, epistemology does not justify our knowledge; instead, it *explains how knowledge is acquired*. It does not need to justify our knowledge because we do not begin with global skepticism, the worry that we could be wrong about everything we think we know. This skepticism is beautifully set out by Descartes in the *First Meditation*. Unlike Descartes, Nelson and Quine begin from the position that our beliefs are true and that we *do* know. The distinct contribution of epistemology is to examine our best cases of knowledge – among these are the results of science in the narrow sense, i.e. the natural and social sciences – to figure out why and how we do in fact know. Naturalized epistemology will explain how we construct theories, including "common-sense theories" that we use every day to get around in the world as well as our scientific theories. So if our theories are successful, i.e. help us make sense of and predict our experiences, then in giving a correct account of how we know, an epistemology describes the norms of successful theorizing. Epistemology describes the successful norms that we do in fact use. Here we can see why it is so important for an epistemology to give an empirically adequate description of how and why we are successful.

Normative questions in philosophy of science will no longer turn on what criteria justify the decisions made by individual scientists; instead, the questions turn on

> whether the social processes that currently characterize scientific practice – those involving the recruitment or education of scientists, for example, peer-review mechanisms, and so on – are the processes that *should* be at work – are likely to produce the *best* theories and research programs – and, if not, what changes should be made to insure that they do.
>
> (Nelson 1996: 103)

We will return to the normative nature of philosophy of science below.

A particular theory or claim (whether common sense, scientific, or philosophical) is justified, in the end, by its ability to make sense of what we experience and to predict our experience. What does Nelson mean when she says that a theory "makes sense" of our experience? She means that theories provide "bridges" between otherwise unconnected sensory experiences. Without theories, an individual's experiences would be a meaningless barrage of sensory stimulations.

1.4 Feminist social epistemology

So how are theories constructed? Here we turn to Nelson's greatest departure from Quine and many other empiricists: she argues that knowledge is social; theories are produced and maintained by communities; therefore, we need a *social epistemology*, SE: "The appropriate loci of philosophical analyses of science are science communities, with the standards, theories, and practices of such communities the appropriate loci of philosophical explanations and evaluations of scientific practice" (Nelson 1996: 101).

Nelson gives many related arguments for SE; here we will present four of them.

1 SE follows from the demise of the theory/observation dichotomy. On the logical empiricist and on other positivist accounts, there is a determinate set of observations from which a hypothesis can be derived, e.g. on Bertrand Russell's account, a scientific claim is bi-conditional upon a list of sense-data, set out in a sense-data language. Later, Hempel's Hypothetico-Deductive Model of Confirmation attempted to set out the logic by which "direct observations" – somehow independent of all theory while still publicly agreed upon – confirm a hypothesis under specified conditions. The problem, as Popper saw, is that there are no such observations, independent of all theory and functioning as the foundations of knowledge. Scientific observations occur using techniques agreed upon by the scientific *community* and the techniques themselves rest upon bodies of theory and are *socially* agreed upon (Nelson 1990: 44–60).

   Scientific observations are set out in sentences using terms with *publicly* agreed upon meanings. And ultimately, such sentences use common-sense, non-scientific terms that are also *publicly* agreed upon. How are we to understand observations? Once foundationalist philosophical explanations that depend upon the theory/observation dichotomy are ruled out, naturalists turn to current scientific explanations of them. Thus, current science tells us that our sensory experiences are to be explained as the firings of sensory receptors in the brain, but Nelson reminds us that "we do not consciously experience those firings." We experience colorful, noisy, smelly things. These sensory experiences of things are made possible by and structured by theories, particularly common-sense theories, which we begin to learn when we are very young and are learning language. And these theories are generated and maintained by our communities. "We experience the world through the lens of the body of theory generated and maintained by our communities" (Nelson 1995: 407). We will examine Nelson's concept of community below.

2   As a matter of empirical fact, language cannot be acquired by an isolated individual. Feminist science has shown the "necessity of socio-linguistic and other environments for the post-natal neurobiological development which permits language acquisition and higher-brain functioning" (Nelson 1995: 408). Language acquisition, conceptual-ization, and many perceptual experiences are, of course, only possible following the development of brain structures, especially "the major expansion in the dendritic and synaptic network, occur[ing] only in complex interaction with the rest of the organism and external stimuli, including, *necessarily*, interpersonal experience" (Nelson 1990: 285). Citing the work of Walker (1981), feminist scientist Ruth Bleier argues for non-linear, causal/functional models of the relationships between postnatal neuronal development and input provided by inter-personal relationships (Bleier 1984 cited in Nelson 1990: 286; see also Hubbard 1982).

Some philosophers have turned to evolutionary accounts of the evolution of human brains to explain how individuals, in principle isolable from one another, have come to be able to have a rich set of concepts, make conceptualized observations, and acquire language. But Nelson points out that accounts of the human brain in evolutionary biology themselves make it clear that the evolution of the human brain was caused in part by *interpersonal relationships* and *social activities*, such as hunting and gathering. Thus, as Nelson argues, it follows that evolu-tionary epistemology – a very popular naturalized epistemology – must focus on human social groups as well as on neuroscience when explaining how knowledge is produced (Nelson 1990: 285–7).

3   SE rests on the historically specific nature of current androcentric and feminist assumptions.

Nelson argues that an adequate account of how knowledge is produced must be able to explain the existence of both androcentric and feminist perspectives in and contributions to science.

Quine's injunction that naturalized philosophy of science be continuous with its subject matter means not only that philosophy of science should use current science(s) to explain science; it also means that philosophical methods and explanations should be judged by, among others, the standards applied to any methods and theories of science, notably explanatory power and empirical success. These are norms of science and so of philosophy of science. Theories in science and so on in philosophy of science help us organize, explain, and predict experience (Nelson 1995: 399–402). And among the first things a naturalized philosophy of science must explain is how people produce knowledge, not only in the sciences, but also in their common-sense theorizing. With Quine, Nelson views common-sense theories in the same way as the theories of science and philosophy; beginning with physical object theory, common-sense theories also organize, explain, and predict experience, just not as rigorously as theories of science. On this view, any group of people who share experiences, e.g. of privilege or of oppression by gender or race, can work out accounts that allow them to explain and predict such experiences. Each group will, thus, share a perspective, some androcentric, some feminist, and will produce explanatory assumptions and theories. These survive to the extent that, like any common-sense theories, they are empirically successful – allowing the explanation and prediction of experience.

And Nelson reminds us that such perspectives are social. "A solipsist," she says, "could be neither a feminist nor androcentric." Not only are *successful* feminist theories produced by people working together, but their content reflects and requires experiences of specific sorts of political and social contexts. "A host of categories, social relations, practices, experiences, and assumptions are necessary for these perspectives to be possible," and all these factors require other people; they are necessarily intersubjective (Nelson 1990: 267). Both androcentric and feminist assumptions, methodologies, models, and theories "have been generated within social experiences, relations, traditions, and historically and culturally specific ways of organizing social life" (Nelson 1993: 147). But we see this most clearly in the rise of feminisms – responses of differently situated women to the social relations of gender, and to the historically and culturally specific ways of organizing gender in which they find themselves. As they strive to understand their experiences of the social world they live in, women (and some men) produce feminist theories of explanation. When these conflict with the assumptions underlying some mainstream scientific models and theories, feminist scientists are well-situated to call into question those assumptions and the results that depend on them.

Thus, Nelson argues that the feminist recognition of androcentrism in some scientific theories must be explained socially (SE). Epistemic individualism explains androcentrism by arguing that the theories at issue are cases of bad science, i.e. the "capacities" of the individual scientist were compromised, e.g. by gender bias. On this account, if we accept the soundness of the feminist criticisms, we are forced to say that the feminists were better scientists than their non-feminist colleagues; however, there is another explanation, viz. that their feminist standpoint, their political views of and experiences of gender relations, were not irrelevant to their scientific insights, but enabled them (Nelson 1990: 268–9).

More deeply still, the changes brought about by feminist science critics show that the *social processes* that brought about the changes are part of the same network that includes *bodies of evidence* in the broad sense required by holism. The beliefs about gender, about proper social arrangements in the larger society and in science, are part of the broad set of theories that ultimately constitute evidence for any single scientific hypothesis. This becomes especially clear when 1) a group of scientists – here feminists – with a particularly salient set of experiences, beliefs, common-sense theories, and values dispute established claims as dependent in part upon different – here androcentric – beliefs, common-sense theories, and values; and 2) the disputants succeed in changing the established claims. Nelson says,

To see this, consider the question of how it was that feminist scientists in a range of fields and disciplines recognized the role of gender and values that their colleagues did not, or why the designation of different strains of E. *coli* as "male" and "female" in a 1986 edition of a widely-used text in molecular biology was dropped in the second edition, or why the following passage, which biologist Scott Gilbert points out was read by most embryologists educated through the 1970s, has recently been dropped from a widely-used embryology textbook:

In all systems that we have considered, maleness means mastery; the Y-chromosome over the X, the medulla over the cortex, androgen over estrogen. So physiologically speaking, there is no justification for believing in the equality of the sexes.

(Nelson 1996: 113)

The androcentric claims cited here were not baseless; they rested upon evidence that included commonly held social assumptions about relations between the sexes. That feminists did not hold them was due in large part to their perspective, a perspective arising, as we have seen, from their shared response to current "social experiences, relations, traditions, and historically and culturally specific ways of organizing social life" (Nelson 1993: 147).

4   SE is shown by empirical evidence for the social nature of science. (See Addelson 1983, Downes 1993, Fuller 1988, Harding 1983, Longino 1990 and 2002, Maffie 1991, Nelson 1990 and 1993, Potter 1993 and 2001, Solomon 1994 and 2001 and Stump 1992.)

Nelson offers a number of case studies showing that scientific knowledge is socially produced (SE) as well as showing the holistic nature of evidence, particularly including the use of socio-political assumptions and theories as part of the evidence (FAE). These include the centrality of sex differences in sociobiology (see Nelson 1995: 415), the discovery of proton structure (see Nelson 1993, *passim*), the "man the hunter" theory of human evolution (see Nelson 1990: 205–12 and *passim*), and the "executive DNA" theory of gene action (see Nelson 1990: 214). Here we will present only one of Nelson's cases, current work in neuroendocrinology showing that hemispheric lateralization is sex-differentiated.

1.5 A case study

Nelson tells us that

On one formulation [of the hypothesis that hemispheric lateralization is sex-differentiated], androgens have an organizing effect on male fetal brains, causing right-hemisphere dominance in the processing of visuo-spatial information, a dominance related in some of the research I will summarize to "superior performance" of males in "spatial contexts" and, among human males, in mathematics. [Nelson uses quotation marks here because the studies she summarizes do not adequately specify what constitutes superior performance or spatial contexts.]

On the basis of studies suggesting correlations among migraine, left-handedness, immune-system disorders, and learning disabilities, and that the last three are more common in males than in females, Geschwind and Behan (1982) proposed that testosterone causes right-hemisphere dominance in males by slowing the development of the left-hemisphere cortex (p. 5099). They cited results in two additional studies as providing evidence for the causal relationship. Diamond et al. (1981) reported that two areas of the cortex of male rat brains are 3 percent thicker on the right side than the left, an asymmetry not found in female rats, and suggested that the thickness was related to lateralization. Appealing to a current hypothesis in empirical psychology – that right-hemisphere lateralization increases visuo-spatial ability – Diamond et al. noted that such lateralization would better enable male rats to interact with female rats during estrus (p. 266). Geschwind and Behan supported their extrapolation of Diamond et al.'s results and hypothesis to humans by appeal to Chi et al. (1977) which reported that two convolutions of the right hemisphere of human fetal brains develop several weeks earlier than do corresponding convolutions of the left. Geschwind and Behan suggested that the differential development Chi et al. reported was caused by testosterone. Two years later, Geschwind and Behan appealed to a study in empirical psychology reporting "a marked excess of males" among mathematically gifted children as additional evidence that testosterone causes right-hemisphere dominance (Geschwind and Behan 1984, p. 221; cf. Kolata 1983).

Feminist biologists critical of this hypothesis argue that there is no theoretical basis for the relationship posited between lateralization and thickness in the right hemisphere of male rat brains. Without this, they contend, there is no basis for the hypothesis that testosterone causes right-hemisphere lateralization. They also maintain that the "higher level" hypothesis of human sex-differentiated lateralization relies on unconfirmed and controversial hypotheses, including the hypotheses that cognitive abilities are sex-differentiated and that such differences (were they to be established) have a biological foundation. And they question the rationale of looking for a biological foundation for the sex differences assumed, given that changes in social expectations and educational policies are closing the gap between girls and boys in mathematical performance (a gap sex-differentiated lateralization was initially proposed to explain) and that a substantial body of research documents significant differences in relevant socialization.

Feminist biologists also claim that results have been misrepresented in some of the studies alleging to explain sex differences. I noted, for example, that Geschwind and Behan (1982) appealed to the differential development of hemispheric convolutions in human fetal brains reported by Chi *et al.* (1977) to support the hypothesis that testosterone causes right-hemisphere dominance. What they failed to state is that Chi *et al.* reported the differential development in both female and male brains, and noted that investigators "could recognize no difference between male and female brains of the same gestational age in the measurement of 507 human fetal brains of 10–44 weeks gestation" (Chi *et al.* 1977, p. 92) – results obviously undermining Geschwind and Behan's claim that the differential development supported their hypothesis. . . .

Considered in isolation, Geschwind and Behan's hypothesis does seem unwarranted. But the methodological principles I advocate lead to a different assessment. According to the first, the evidence for a hypothesis or research program includes both direct empirical success (the successful prediction and explanation of relevant data) and how the hypothesis or program is integrated into other accepted theories and research, inclusive of those informed by or expressive of values and sociopolitical assumptions. By this standard, there was evidential warrant in the 1970s and at least early 1980s for Geschwind and Behan's hypothesis. . . .

Neuroendocrinology's central research questions involve relationships among hormones, neural events, and behavior. Given the assumption of "male" hormones and "female" hormones (androgens and estrogens, respectively), and a broader assumption of sexual dimorphism, males and females provide a "natural" base line for investigating these relationships. Moreover, research built on a male/female dichotomy was supported by hypotheses and experimental results in closely allied disciplines. At least 17 years prior to Geschwind and Behan's hypothesis, investigators in reproductive endocrinology reported that androgens block the cyclical response of hypothalamic neurons which regulate pituitary functions in female rats. While the effects of estrogens received far less attention in these investigations, the blocking effects were taken as evidence that androgens "organize" hypothalamic neurons such that rat brains become sexed i.e., that androgens organize "a male brain" (Harris and Levine 1965). In the intervening years, studies in empirical psychology and endocrinology (most involving rats) suggested causal relationships between androgens and "cognitive capacities" (e.g., maze-negotiating abilities) and behavior (e.g., "aggression", defined and measured on the basis of fighting encounters among laboratory animals, and what were designated as "normal male sexual behaviors").

Links among these several lines of investigation, and research into cerebral dominance in neuroanatomy, were forged four years prior to Geschwind and Behan's hypothesis that testosterone causes right-hemisphere lateralization. Reproductive endocrinologists reported morphological sex differences in the brain of some bird species in areas previously related to the ability of males to sing, and sex differences in the number and size of neurons in areas of rodent brains associated with the regulation of estrous cyclicity (Gorski et al. 1978 and Gorski 1979). Gorski et al. 1978 announced that "The concept of the sexual differentiation of brain function is now well established" and called for further investigation into its mechanisms (p. 334); Gorski 1979 reported the search for "a clear morphological signature of sexual differentiation in the brain" (p. 114). Finally, evidential warrant for Geschwind and Behan's hypothesis was provided by the investigations in empirical psychology into sex differences in lateralization and the effects of lateralization on cognitive abilities; by studies (of which I have mentioned only one) claiming to establish "clear sex differences" in mathematical abilities; and by apparent correlations between testosterone and left-handedness, immune-system disorders, and learning disabilities.

When evidence is construed holistically [FAE], Geschwind and Behan's hypothesis is revealed to be neither far fetched nor purely politically motivated. It represented the synthesis of central research questions, current hypotheses, and experimental results in three research traditions, and the convergence of these with cultural assumptions about sex differences (e.g., that males have superior spatial and mathematical abilities, that there is a biological foundation for these sex differences, etc.). . . . The hypothesis of male right-hemisphere dominance is not plausibly written off as "bad science." There was substantial evidence for it, constituted by research traditions and experimental results in several disciplines, and widely accepted assumptions about sex differences.

(Nelson 1995: 410–13 and 414)

This case shows us the *social* nature of both mainstream and feminist scientific knowledge (SE):

Focusing on the communities involved, as the second methodological principle [SE] calls for, reveals that this hypothesis was commensurate with research questions central to neuroendocrinology and closely allied disciplines. Endocrinologists and empirical psychologists frequently cited and built on each other's hypotheses and research results in the investigations just summarized, and, as early as the 1970s, the study of relationships between gonadal hormones, brain structure, cognitive capacities, and behavior involved collaborations across these disciplines, and neurophysiology and neuroanatomy (see, e.g., Geschwind and Galaburda 1984, *Science* 211 (1981), and Nelson and Nelson 1996).

But the criticisms feminist scientists have leveled against this hypothesis reveal that aspects of the research just outlined were problematic and characterized by a fair dose of androcentrism. Feminist biologists point to the emphasis on "male" hormones in these investigations and disciplines, and argue that conclusions cannot be drawn on the basis of research apparently establishing their organizing and activating effects until a similar amount of research is devoted to the effects of estrogen. Some also challenge the labeling of androgens and estrogens as "male" and "female" (and the broader assumption of sexual dimorphism in these disciplines), noting that males and females produce both; that the circulating form of progesterone (which is metabolized to testosterone, the major androgen) is also metabolized to estradiol (the major estrogen); and that, among the three families of sex hormones, there are continuous conversions of some forms to others. Feminist scientists also cite experimental results that indicate complex and often non-linear interactions between cells, and between cells and the maternal and external environments, during every stage of fetal development, and experimental results involving other species (e.g. guinea pigs) that conflict with those involving rats. On the basis of these several arguments, they maintain that the isolation of so-called "male" hormones as causing fairly remote effects is unwarranted. Finally, feminist scientists point to statistical analyses that indicate more variation among members of each sex on standardized tests than between them, and that establishing a sex-difference in cognitive abilities would require that more variables be taken into account than have been to date (e.g. Bleier 1984, 1988).

> The evidence for these critiques, like that for the hypothesis with which we began, encompasses a broad body of experimental results, current hypotheses, and theory – elements of which, like the research feminists criticize, are substantively informed by sociopolitical context and values. The recognition of the role of androcentrism and assumptions about gender in the original research – in shaping the objects for which explanations are sought, the descriptions of behavior of laboratory animals and humans, the assumption of sexual dimorphism, and the lack of attention paid to estrogens and to alternative, sociological explanations of assumed sex differences – attests to the integration of results, standards, theories, and research backgrounds of scientific disciplines and feminist communities [FAE and SE]. . . . It is as clear that neither the hypothesis Geschwind and Behan advanced, nor feminist critiques of it, are plausibly construed as the product of any individual's efforts (or even those of several). Each represents the synthesis of theories maintained, methods accepted, research undertaken, and results achieved by communities [SE].
>
> (Nelson 1995: 413–14)

This case is one of many showing us that scientific *communities and sub-communities* produce knowledge. Several scientific sub-communities use one another's results in ways that reveal the network-like nature of their many different theories and, thus, the communal nature of those theories. We also see that feminist and non-feminist scientists belong to a larger research community, sharing many, many theories, methods, and standards – all of which are publicly agreed upon. Moreover, the feminist science critiques arise from the shared experiences and shared understandings that typify the social and political communities of which feminists are a part; these experiences, understandings, and communities enabled the critiques. (Finally, the case is one in which values informed the original, androcentric research without making it bad science.)

1.6 Epistemic communities

The answer to the question suggested by the title of Nelson's first book, *Who Knows*, is that "we know." And the "we" refers to an epistemic community of which "we" are a part. What is an epistemic community and how do social epistemologists determine the boundaries of epistemic communities? Nelson argues that what counts as a community will be case specific; thus, a scientific community is a group of scientists who share research questions and research traditions, standards, theories, and methods – they may share all of these or enough of each that we can say they are working in the "same research area or field" even though there may be some disagreement within the community. In some cases, the community will be a narrow scientific discipline, while in others the community will cross traditional disciplinary or scientific boundaries. Sociologists of science often use empirical factors to determine who belongs to a research community or area; these include institutional factors such as departments, professional associations, reading or publishing in the same journals, attending the same conferences or conference sessions, collaborating on research projects, using one another's research results and citing each other in pre-prints and articles (Nelson 1996: 103). Nelson does not think there exist any a priori criteria for determining the boundaries of an epistemic community; here, as a naturalizing philosopher of science, she turns to social sciences such as sociology and anthropology to make reasonable determinations on a case-by-case basis.

Scientific and other knowledge-producing communities have dynamic, fuzzy boundaries. Setting boundaries, Nelson says, depends on our purposes, e.g. doing epistemology, forming academic sub-communities, political action groups, neighborhood groups for local issues, etc. None of these epistemic communities is monolithic or stable. They are not monolithic because each member accepts some of the knowledge, standards, and categories that belong to the group as a whole, but not all agree on everything and there may be no single belief that is held by all, e.g. feminists constitute such a group; feminists have found it important at this time in history to recognize the many differences among them (Nelson 1993: 149–50).

But communities do not dissolve into collections of knowing individuals, each in principle able to know without the others. Our individual experiences and the fact that we each belong to more than one epistemic community enable each of us to make unique contributions to the knowledge produced in our many communities. Someone may have an experience that no one else has had, but what she knows "on the basis of that experience has been made possible and is compatible with the standards and knowledge of one or more communities" of which she is a member. One cannot make a contribution to knowledge without public standards that determine what one's community will recognize as knowledge; indeed, it is public standards and knowledge that enable one to organize one's experiences into coherent accounts in the first place. So while individuals put forward candidates for knowledge, "none of us knows (or could) what no one else could" (Nelson 1993: 149–50).

1.7 Epistemic individualism

To help us understand what Nelson means when she says that "none of us knows (or could) what no one else could" and that communities are the appropriate loci of the analysis of science, we must understand what she is denying. She is denying the general picture according to which autonomous individuals can produce knowledge and the particular version of that picture as it applies to science, viz. that individual scientists autonomously produce scientific knowledge. Sometimes the general picture is spelled out as solipsism, e.g. as the rationalist Descartes presents it in his *Meditations*. But empiricists have also presented solipsistic accounts of how knowledge is produced. For example, Locke tells us that the (infant) human mind is a blank slate, equipped with certain *abilities* but without any ideas. The first ideas to come into the mind are sensory ideas that are caused in the mind by the presentations of the five senses and are representations of those presentations. Using only his innate abilities, among which the most important are the abilities to abstract and compound ideas, the epistemically isolated, autonomous individual has ideas representing the world around him to which he attaches words when he learns language. Book III of Locke's *Essay Concerning Human Understanding* tells us how language is built up in this way, and Book IV tells us how knowledge is built up using abstracted, compounded, and other ideas all derived ultimately from sensory ideas. Locke's epistemology is a good example of epistemic individualism, for although he agrees that other people exist, he argues that each of us is what Putnam once referred to as a "methodological solipsist." To grasp the public language he is being taught, the individual must do the work in the way described by solipsism (see Putnam 1983).

Nelson explains the allure of epistemic individualism for empiricists as, in part, due to the demand that our theories (of which scientific theories are paradigm examples) be tied to the world. For Quine, the only evidence for our hypotheses and theories, or for any of our beliefs, comes to us through our sensory receptors. But sensory receptors are features of individuals, so it is natural to take the individual as the locus of knowledge. And as naturalizing epistemologists, Quineans use empirical psychology and evolutionary biology for a partial account of how we get a wealth of hypotheses, theories, and research programs from the mere firings of our sensory receptors (Nelson 1995: 402 and 1990: 276). But as we have seen, naturalism does not require epistemic individualism; in fact, empirical studies of how scientific hypotheses and theories are produced reveal that they are produced by groups of people, i.e. by communities (demarcated in the ways we mentioned above) and that the development of the individual human brain, which allows the individual to acquire language and perform other higher cognitive functions, requires a "socio-linguistic environment", i.e. other people who are speaking a language.

Moreover, the connection between our rich set of theories about the world and the world itself is a connection between our theories and our experiences of the world. These experiences are "made possible because we have sensory receptors and are able to learn public theories," and it is in terms of these public theories that we experience the world. We do not, as the positivist and post-logical positivist empiricists had hoped, have untheorized, "raw" experiences or observations. Instead, we experience the world through "the conceptual scheme, the theory of nature, we begin learning as we learn language." She agrees with Quine that sensory experiences are coherent only because they are "bridged" or connected to one another, shaped by public language and a conceptual scheme. Because there is no coherent pre-theoretical experience and because the theories that bridge our experiences are produced socially, a solipsistic knower is impossible. The child depends epistemologically on an evolving public conceptual scheme to bridge her or his sensory experiences, connecting them, categorizing and "shaping" them. And in keeping with the demands of empiricism, these experiences constrain what it is reasonable to believe, but they do so only together with our going, public theories (Nelson 1990: 276, 288).

1.8 Who knows?

Nelson does not deny that individuals know things; rather, she denies that individuals know anything *autonomously, in epistemic isolation*; she denies "the view that persons – complete with projects, characteristics, beliefs, values, capacities, and interests – are, at least in principle, separable from society. That is, persons have such characteristics without essential need of social context or others" (Nelson 1990: 257). Her view is that persons cannot have beliefs, values, interests, etc. without interacting with other people in a socio-linguistic environment. Here she cites Alison Jaggar's important insights that "Human interdependence is . . . necessitated by human biology, and the assumption of individual self-sufficiency is plausible only if one ignores human biology" (Nelson 1990: 257; Jaggar 1983: 40–1).

Although for Quine "Mama" is the name of an object rather than a subject, Nelson agrees with feminist object-relations theorist Nancy Chodorow (1978) that while the child is learning to individuate objects, she or he is simultaneously learning to individuate the self and learning about other persons, especially the primary caretaker – usually the mother. And the most important factor in a child's recognizing itself as an individual subject is coming to see the mother as a subject. Thus, the child is learning far more than how to be a member of a linguistic community, but how to relate to others and be a member of a human community. Nelson says,

It is a consequence of Quine's own arguments against foundationalism, including his account of how a child acquires language and public theory, that we do not develop epistemologically . . . without intersubjective experience, and that we do not, individually or collectively, observe or experience the world from outside the theories our species and our particular communities have developed to date.

(Nelson 1990: 289–90)

We need public methods, standards, and theories to determine when we know something; a scientist can be said to know the hypothesis she thinks up only after it has met the standards of evidence of the scientific community of which she is a part. But we also need public theories to make sense of our experiences and to have beliefs in the first place. Nelson says,

What constitutes evidence for a claim is not determined by individuals, but by the standards a community accepts *concomitantly with* constructing, adopting, and refining theories. Those standards constrain what it is possible for an individual to believe as well as the theorizing we engage in together.

(Nelson 1990: 277)

Thus, she argues that *autonomous* individuals not only cannot have knowledge, they cannot have beliefs, either. Only members of communities can have *reasonable* beliefs. Here reasonable belief is contrasted with irrational, bizarre, or meaningless beliefs. Standards of rationality and of sense and nonsense, related to and including many standards of evidence, are set by the community, not by doxastically autonomous individuals. The theories that in part constitute our experiences – the ultimate grounds of our beliefs and of evidence – are the product of social activity and individual experience, and an individual's beliefs are impossible without them. An individual scientist is able to think up a hypothesis because she is already immersed in a very large system of common-sense and scientific theories that make the hypothesis meaningful and reasonable (if it is). An individual's beliefs depend on the standards and theories of the community in order to be relevant to, bear on, properly apply to, or be pertinent to the matter at hand. Individuals work in research traditions that have accepted theories and methods, so when an individual produces a hypothesis, she or he is not independent of others in the tradition and in the community (Nelson 1990: 277 and 1995: 409).

1.9 Relativism

If the standards and methods by which knowledge is determined are not stable over time and so evidence is historically specific, and if epistemic communities and sub-communities differ in their perspectives and these perspectives can find their way into their hypotheses and theories, and even into their methods and standards, doesn't relativism necessarily follow?

What kind of relativism is supposedly a threat to us? We can distinguish many relativisms, e.g. (R_1) the relativism of individual claims to knowledge, whereby we are supposed to be unable to decide rationally among the conflicting claims put forward by autonomous individuals; (R_2) the relativism of different community claims to knowledge, whereby we are supposed to be unable to decide rationally among the conflicting claims put forward by communities; and (R_3) what we might call "global relativism," which we will briefly discuss below. (These are not the only ways to distinguish among relativisms.)

As we have seen, Nelson denies that individuals know autonomously, i.e. she denies epistemic individualism; but she also denies that this entails relativism (R_1) because knowledge is produced by communities, and using the public standards and theories of the community, we can distinguish among the claims put forward by individuals. Turning to the charge that on Nelson's account we cannot distinguish on epistemological grounds among the competing claims of different communities (R_2), let us take scientific communities (in the narrow sense) as our example. Specifically, if standards of evidence are relative to the times and to communities, i.e. arise in part because of the specific social relations that affect the social processes through which communities determine those standards, determine what the evidence is on the basis of them and so determine what hypotheses and theories to adopt, isn't Nelson committed to the view that (R_2) there is no way to distinguish between the conflicting knowledge claims put forward by two different communities?

Science is an ongoing enterprise and its standards and theories are certainly historically relative, but historical relativism, particularly the historical relativism of community standards and so of evidence, need not entail judgmental (sometimes referred to as "epistemological") relativism – R_1, R_2 or R_3. This is because, as an empiricist, Nelson assumes that experience is the final arbiter of claims and theories, and that the goal of science is successful explanation and prediction of those experiences. Thus, *evidential warrant* is central to her account of scientific (indeed, any) knowledge and it is not the case that all theories or claims are equally compatible with experience. If, as Kuhn argued, disputing groups of scientists inhabited incommensurable paradigms, then there would be no way for them to determine which of their different methods and assumptions were ultimately warranted by the evidence. In her insightful discussion of Kuhn's notion of the incommensurability of scientific paradigms, Nelson points to his hidden assumption that scientific communities are not only cut off from one another, but also cut off from the common-sense communities within which sciences operate. The result is, of course, that there is no way for anyone to decide between (allegedly) conflicting paradigms on the grounds of evidence and/or other criteria of theory choice; such "choices" would have to be made on emotional or other non-scientific grounds best explained by psychology or sociology.

But Nelson argues that, while there are certainly disagreements between sub-communities of larger scientific communities and, sometimes, between scientific communities, these communities all share many common-sense beliefs and theories. For example, advocates of the "man the hunter" view of human evolution and their feminist critics disagree about models and observations, and perhaps about some common-sense assumptions and knowledge about gender. But they share a larger body of knowledge and standards, e.g. physical object theory and the view that humans evolved and that their activities were factors in the evolutionary process. "Hence, members of these groups can discuss (and disagree about) the significance of 'chipped stones' without any lapse in conversation" and they can use knowledge and standards they share to evaluate conflicting claims. They can communicate and can and do, in fact, understand their points of agreement and disagreement (Nelson 1993: 147–8). Nor is epistemological relativism entailed by the fact that scientists might go on disagreeing for a very, very long time, perhaps until humans die out.

At this point, critics charge that Nelson's response misses the scope of the problem. Relativism is not a threat when disputants can agree to disagree, i.e. they agree on a sufficient number of beliefs and methods and standards to figure out their points of disagreement. (In the worst such case, they agree that evidence should decide their dispute, but disagree about what would count as evidence that would decide their dispute.) No, the real threat of relativism arises when (R_3) there is, in principle, no way to decide between complete systems of knowledge, theories, methods, standards, and goals. (This is a version of Kuhn's incommensurability of paradigms.)

Nelson notes Quine's assumption that "we" all share the same theories, methods, standards, and so on. But Quine's underdetermination thesis says that for every hypothesis and theory there is in principle an empirically equivalent hypothesis or theory, and he argues that since a hypothesis does not face observations alone, but does so along with all our going theories, then we can maintain a hypothesis in the face of recalcitrant observations if we are willing to change the rest of the system accordingly, i.e. redistribute truth values among all the sentences constituting the network of our knowledge. From this it follows that we can imagine (at least) two different systems of knowledge both accounting for the evidence provided by our observations. The systems are empirically equivalent, so they cannot be distinguished on the basis of evidence. And this global relativism (R_3) is pernicious.

Nelson does not discuss global relativism, but it is consistent with her philosophy of science to argue that her view is an empiricist one according to which knowledge in general and scientific knowledge in particular provides us humans with understandings of the world that enable us to explain and predict our experiences and so to negotiate our way about in our world successfully. Most notably, Nelson does not hold the rationalist view that *the purpose of knowledge and of science is to discover truth*. (Though she certainly does not deny that hypotheses and theories can be, and mostly are, true.) Her view may, therefore, be said to be a feminist pragmatism according to which the purpose of science is to allow us to negotiate our way about in our world successfully using empirically successful, predictive theories. Hence the threat of relativism is that it will disenable us from successfully negotiating our way about in our world. But the relativism of global systems, each of which is empirically adequate, does not entail that we are not empirically successful, negotiating about in our world, taking our lumps when our hypotheses are unwarranted and our pleasures when they are borne out by our experience. As long as both systems allow us to be empirically successful, both are *epistemically* tenable. R_3, then, is not pernicious for feminist pragmatism.

1.10 Norms and naturalizing

With other feminist philosophers of science, Nelson points out that contemporary scientific communities are homogeneous enough that they hold many background assumptions, e.g. beliefs about gender, which remain invisible to their members. Heterogeneous science communities, on the other hand, are more likely to hold heterogeneous background assumptions, and so are more likely to recognize otherwise unnoticed, taken-for-granted assumptions and facilitate critical examination of them (Nelson and Nelson 1994: 494).

Now what is the status of this point? Is it an empirical or a normative point? Nelson is *suggesting* that scientific communities be more heterogeneous so that background assumptions, e.g. about gender, will more likely be examined. Can naturalized philosophies of science be normative after all, evaluating existing standards and methods, and suggesting improvements?

The division between normative, on the one hand, and empirical or descriptive, on the other, itself depends upon a sharp distinction between facts and values, thought to be reflected in the sharp distinction between the empirical findings of science and the value claims of moral, religious, or other value systems. But it is this division and this sharp distinction that Nelson's holism denies. With Quine, she advocates a holistic view of knowledge according to which there are no sharp boundaries among common sense, science and philosophy, but, as we have seen, unlike Quine, she advocates a holistic view of evidence (FAE) according to which evidence for any hypothesis includes not only experience but also theories (in the natural and social sciences, philosophy and common sense), including those "informed by social beliefs and values." It follows from Nelson's holism that science, philosophy, and common sense are not epistemically sealed off from one another. These are all theories attempting to make sense of our experience; they are all subject to requirements of empirical adequacy. Moreover, results in one field can be and are taken up in the others. It follows that values are not so different from facts; values are subject to modification in light of facts and facts are not "value-free."

We are now in a position to see why Nelson denies the traditional philosophical distinction between a "context of discovery" – the actual circumstances under which a hypothesis or model is conceived, circumstances which might be heavily laced with values, political aims, or what have you – and a "context of justification" – which is supposed to include only the carefully controlled circumstances in which observations are made in order to confirm or disconfirm a hypothesis or model. This distinction was first made by Reichenbach (1938) because he recognized that cases of real scientific work do not fit his elegant logic of science and must be rationally reconstructed to fit it. The distinction was taken up by positivist and other philosophers in ways that insured the irrelevance of social and other values to the rational acceptance of scientific beliefs. Thus, the distinction rests in part upon the *assumption* of a sharp distinction between moral and social evaluation, and the justification of scientific results.

It is a consequence of Nelson's rigorous holism that she can consistently put forward a normative suggestion for the improvement of science, speaking *as a philosopher of science* or simply *as a member of the common-sense community* to which the sciences are connected in the holistic network of beliefs, a large chunk of which are part of the evidence for scientific beliefs. We should also note that it follows from holism that philosophy of science, like science, is normative. Assuming that the primary goal of science and philosophy of science is to make theories that organize, explain, and predict experience, then their results are evaluated by how effectively they do these things (Nelson 1996: 98). Thus, many norms can be put forward in both of these areas but they will ultimately contribute to and rest upon norms for empirical success (see Solomon 2001).

We will return to the intersections of science and values in Chapter 3, setting out Anderson's argument that research work justifying hypotheses in the sciences must in many cases appeal to social values and that, if we are reasonable people, our values are not impervious to the facts; thus, the findings of science shape our values.

2

FEMINIST TACKING BETWEEN SCIENCE
AND PHILOSOPHY OF SCIENCE

Holism facilitates one of the central goals of feminist philosophies of science, viz. finding ways to capture the relations among feminist, sexist, and androcentric common-sense beliefs and the beliefs or hypotheses constituting scientific theories; thus, certain models of holism comport well with both Wylie's version and Nelson's version of naturalized feminist empiricism.

Quine's way of modeling holism was to suggest a "web of belief." The sentences expressing the beliefs or hypotheses constituting theories in ~the natural sciences are logically related to one another and also to the sentences of the social sciences, to those of philosophy and to those of "common sense." Some critics object to this model because in it all beliefs are related to and confirm one another in the same way with the result that it offers a viciously circular account of confirmation. This objectionable model of holism is not one to which holists are necessarily committed. In Section 4 below, we will see that Wylie's Consilience Model of Confirmation gives us the details of a different understanding of holism in which the confirmation of scientific beliefs is not viciously circular.

2.1 Feminist philosophy of archaeology

Although archaeology is understood to be a social science because it takes as its subject domain the material culture of people who lived in the past, usually before people left written records, it is in fact a hybrid science, using specialized areas of the physical sciences, e.g. for carbon dating; the life sciences such as paleobiology, e.g. for understanding human and animal remains; paleobotany, e.g. for determining the plants people used and ate; and paleoenvironmental sciences, e.g. for determining the environmental conditions under which people lived in the past. These and other physical and life sciences help archaeologists interpret their data so that it can be used as evidence, along with evidence provided by social sciences, including, for example, ethnohistory and ethnography, to test hypotheses and theories about past cultures.

A challenging problem for archaeologists has been confirming hypotheses about the lifeways of people who have been dead for hundreds or thousands of years and about whom we have no historical record. How, for example, did they organize their societies? A mere glance at descriptions of how different people living now organize their societies shows such a great variety that it seems very difficult indeed to figure out, on the basis of a few enigmatic physical clues, anything about people living very long ago. What were the gender relations among them? How can we figure out their gender relations when it is sometimes difficult to figure out which skeletal remains are female and which male?! Did they have anything that we would recognize as rituals, religions, etc.? If so, did women take important roles in them? Because they face such great difficulties, archaeologists have been among the most self-critical of scientists. And they have been creative and very careful in constructing a science that can answer our questions in credible ways. Needless to say, this science is very lively, full of disagreements about the best ways to proceed; and it has had closer working relationships with philosophers of science than many other physical and social sciences.

Among the philosophers who have worked with archaeologists over the years is Wylie who, early on in her career, did archaeological field-work at sites in Canada, producing and helping to produce field reports; and she has continued working with archaeologists on the meta-archaeo-logical and ethical problems facing them as they try to understand the human past. In the 1980s, Wylie worked out an account of confirmation, of how researchers confirm or disconfirm their hypotheses and the theories giving rise to them, designed to capture the complex work archaeologists do when they argue from the fragmentary material record uncovered in archaeological work to models that attempt to reconstruct the lifeways of people in the prehistoric past. A pressing problem here is the threat of circularity; in an archaeological context, roughly speaking, this is the threat that arises because researchers cannot infer directly from the material record to the lifeways of prehistoric people; they must "tack" from the present to the past, usually arguing indirectly from us to another "them" with whom we are contemporaries or about whom we have historical information, and thence to "them" in the prehistoric past. Archaeologists cannot, of course, just assume that lifeways in the prehistoric past were similar to anyone's recent lifeways.

2.2 Understanding "them": circularity and ethnographic tacking

The particular form of circularity that has exercised archaeologists (just as it has other scientists) arises from the theory-ladenness of the evidence they use. In the worst case, the evidence is laden with (that is, interpreted in the light of) the very theory giving rise to the hypothesis or model it is being used to test. To see how this problem arises, let us begin with the ethnographer's work; she is trying to understand a group of people whose language and lifeways are unfamiliar to her. Anthropologist Clifford Geertz (1976) described the work she must do as "tacking"; she must take "experience-distant" concepts from her own culture and use them to try to understand the experience-near concepts of the people she is studying. Experience-near concepts are the ordinary, more concrete ones people use as they experience and negotiate the world around them; these concepts are relatively less abstract than experience-distant concepts, such as "person," used to understand other peoples' or one's own experience-near concepts. According to Geertz, ethnographers take distant concepts – theoretical, abstract concepts – from their own culture and tack back and forth between them and the subjects' experience-near concepts – concrete, experience-embedded concepts. This is diagonal tacking across contexts: abstract to concrete, familiar to alien. Wylie developed Geertz's metaphor into an expanded metaphor or model that has been useful for archaeologists. In the first place, the researcher's own experience-distant concepts come from vertical tacking, reflecting on her experience-near concepts and practices, i.e. to uncover her experience-distant concepts, she must understand her own context. Moreover, the ethnographer's subjects have conceptual schemes of their own that order and explain their cultural practices, so they, too, do their own vertical tacking. This means that ethnographers must understand the experience-distant self-understanding as well as the experience-near or experience-embedded concepts of their subjects. Both sets of concepts must be explained by the ethnographer's experience-distant concepts and to the extent that her understanding of the subjects' conceptual scheme conflicts with the subjects' own self-understanding, they are rival explanations of the subjects' lifeways! Ultimately, the challenge for the ethnographer is to adjudicate between the two understandings without circularity, i.e. without presupposing her own account drawn from her own culture's conceptual scheme (Wylie 2002: 164).

We see, then, that ethnographic tacking usually begins with analogical inferences in the generation of hypotheses about other people. To understand others, the ethnographers begin with their own practical knowledge and general theories about human motivations, beliefs, and capabilities. These general theories are thus comprised of their own experience-distant explanatory concepts and are used to frame hypotheses about other people. These hypotheses and theories, Wylie argues, must be tested against evidence that ethnographers have grasped the meaning of their subjects' practices. The ethnographers' model explaining their subjects' lifeways must be responsive to evidence of its explanatory and empirical adequacy or inadequacy.

And, as we would expect, ethnographers and subjects learn from each other; they are engaged in a dialectical process of exchange. Ethnographers do not just grasp the conceptual scheme of the subject community; they might also rethink their own experience-distant concepts and experience-near beliefs and practices. Both can change; and importantly for this science, the ethnographers' criteria of adequacy for what determines a better account of her own and of the subjects' lifeways can also change.

To summarize, Wylie's tacking model makes clear that researchers get their (experience-distant) analytical concepts by vertical tacking between their own experiences and their explanations of them. And these concepts are their initial guide for grasping the experience-distant and experience-near concepts informing unfamiliar practices, i.e. for horizontal tacking between our explanations of unfamiliar subjects and their own explanations of themselves and for diagonal tacking between our explanations and their practices. Each requires an argument to show that we have the right understanding of ourselves, of their practices, and of their self-understanding, and each of our arguments is evaluated by criteria that are the same as those for judging between competing theories within a research tradition or within a conceptual scheme.

Unlike ethnographers, archaeologists do not usually have direct access to beliefs expressed by subjects; their subjects are long dead and left no written history (though they may have left a symbolic record). So these scientists often make explicit the assumptions and inferential steps often suppressed by ethnographers who negotiate directly with people whose lifeways are unfamiliar to them (although, arguably, ethnographers do not have direct access to the beliefs of people in unfamiliar contexts, either).

2.3 Background assumptions

When archaeologists approach the material remains of past people, they want to reconstruct how, when, by whom, or as a consequence of what type of cultural conditions this material record was produced. Some also want to know about the lifeways of the people who left the material record, about the underlying large-scale cultural processes that were manifest in those particulars, etc. When they ask any of these questions, archaeologists use background information from ethnohistory, sociology, psychology, the life sciences, and natural sciences, including ecology, as the basis for their reconstructive inferences. This background information is all part of the archaeologists' contemporary context and serves as the *source* for developing experience-distant theories about cultural development, differentiation, interaction, and adaptation, and for developing experience-near models of cultural practice (vertical tacking). And there are many empirical and conceptual constraints on these vertical, source-side arguments, i.e. constraints on the arguments for the archeologist's experience-distant (analytic) concepts accounting for her culture's own experience-near concepts. The archaeologists also tack from these analytic concepts and interpretive principles to the observable, material consequences of past practices contained in the site(s).

For example, Hill (1966) used information based on the size and shape of pueblo rooms in *contemporary* pueblos and those described in *ethnohistoric* sources to ascribe specific functions to *prehistoric* pueblo rooms in the US Southwest because they were of similar size and shape. And building on Hill's work, W.A. Longacre (1966, 1968) used the ethnohistoric observation that pueblo women potters often work together in family-defined workshops, sharing and influencing one another's designs, to argue that the record – which showed distinctive sets of ceramic designs found on shards deposited at about the same time in spatially different areas of prehistoric pueblos – should be interpreted to mean that clusters of these rooms represented social units – extended, matrilocal and matrilineal family units. These arguments compare material residues of practice in contemporary and ethnohistorical source contexts with residues of practice in a prehistoric context in order to argue for a hypothesis about the social organization of prehistoric practice.

Longacre's interpretation depends on background assumptions about (1) how elements of ceramic design diffuse among potters, viz. smallest units of design are passed on to potters in the same work area, and (2) how a specific social organization of ceramic production, viz. matrilocal work areas, might enhance or curtail this diffusion, producing a distinctive distribution of group-specific styles (Wylie 2002: 166).

Archaeologists also use collateral background knowledge to challenge some interpretive options and establish the credibility of others; for example, Margaret Hardin (1970) challenged Longacre's presupposition that social proximity, i.e. being together in the same workshop, led to sharing design elements. Hardin studies communication among contemporary Mexican potters and has shown that, among these people, the smallest constituents of ceramic design diffuse most quickly and widely. Thus, the designs Longacre used as evidence might have been learned from pottery made at far-away pueblos. Based on her knowledge of contemporary Mexican potters, Hardin argued that it is the overall design structure – into which the small (widely shared) elements are placed – that reflects the close association of potters who work together. From this example, we see how background knowledge is used to support and to challenge archaeologists' claims.

Wylie is a philosophical naturalist in the sense that she wants an empirically adequate model of confirmation; therefore, she investigates how archaeologists in fact judge the relative credibility of claims that something is evidence and the relative credibility of competing interpretive and explanatory claims made on the basis of the evidence. She has found many points of agreement among archaeologists (specifically, among processualists and post-/anti-processualists, who disagree about almost everything else):

1 data and evidence are not given, stable, or independent of theory;
2 identification of archaeological data, i.e. deciding which bits of recovered material to count as data, and their constitution as evidence, depend on "linking principles" (source-side or background knowledge, middle-range theory or mediating interpretive principles); and
3 archaeological data and evidence are interpretive constructs, but, still, archaeological interpretation is not necessarily viciously circular. Depending on linking principles does not guarantee resulting evidence will conform to expectations (Wylie 2002: 174).

For example, although archaeologists can depend on some data as relatively stable, e.g. the size and shape of pueblo rooms, Wylie points out that there are no factual givens or context-neutral reasons that can stabilize interpretation of this sort of data as archaeologists try to determine prehistoric practices and cultural beliefs. Archeological data – facts of the record – are constituted as evidence by "ladening" them with theory, e.g. pottery shards found in pueblo rooms were laden with the theoretical assumption that the smallest units of ceramic design found on them are significant and were passed from mothers to daughters in the same workshop. Nevertheless, the concepts archaeologists start with do not determine what they will find when they tack from source to subject. In this way, they avoid the threat of vicious circularity. Their orienting concepts may be significantly changed by vertical tacks – as Longacre's concepts had to be changed when Hardin showed that the analogy should be made with contemporary potters whose overall design structures are shared within a workshop while their smallest design elements are picked up from far and wide (Wylie 2002: 167 and 173).

Recognition that evidence, even though it must be theory-laden in order to be evidence, constrains acceptable archaeological hypotheses leads Wylie to adopt a philosophical stance of "mitigated objectivism" or moderate scientific realism, which she finds among archaeologists themselves. They see (mitigated) objectivity as a regulative ideal crucial to archaeological practice. They make judgments about the relative credibility of claims about the evidential significance of archaeological data, and while these judgments are not certain, they are not entirely arbitrary either. Wylie recognizes that some archaeologists, whom she dubs "narrowly empiricist objectivists" (we might call them arch-realists), think empirical constraints provide stable, unitary ground for context-independent, objective knowledge. But this view is too simple as we can see from the pueblo pottery example. On the other hand, extreme relativism is also wrong; archaeologists cannot and do not ignore the role of empirical constraints or assimilate them to a seamless and self-contained web of belief by assuming they are arbitrary artifacts of the concepts used to interpret them as evidence.

How can this be? Why aren't archaeologists' claims just arbitrary artifacts of the concepts used to interpret them as evidence? Why aren't they viciously circular? There is a danger that since the analytical theories drawn from familiar sources are used to generate hypotheses about past cultures and lifeways, *and* are also used to interpret archaeological data as evidence to test these hypotheses, then the hypotheses will turn out to be confirmed *because* the theory to be confirmed informs not only the hypotheses used to test the theory, but also the evidence used to test the hypotheses!

2.4 Confirmation of archaeological theories: a Consilience Model of Confirmation

Wylie's Consilience Model of Confirmation is designed to show that and why archaeological reasoning is not circular. She argues that, most often, questions about the adequacy of an interpretive hypothesis are settled when independently constituted lines of evidence converge. Evidence is not a stable, foundational given; it is always "theory-laden," but mitigated objectivity is achieved when the ladening theories – middle-range, linking principles – used to construe/interpret data as evidence are *secure* and *independent*. Her model shows that archaeologists use an enormous diversity of evidence and the diversity ensures that the evidence can sometimes function as a *semi-autonomous* constraint on claims about the cultural past, particularly when some of it depends on background knowledge from one or more different sources and when it enters interpretation at different points. Thus different lines of evidence can be mutually constraining when they converge or fail to converge on a *coherent* account of a particular past context.

In Wylie's model, archaeological evidence gets stability and autonomy, i.e. it is good evidence, if the evidence is *secure* and *independent*. It must be secure in two senses:

(S_1) the background knowledge used to link the present record (data) with antecedent causes (conditions that produced it) or past events must be credible in its home context (e.g. paleobotanical claims used in an argument must be well established in paleobotany), and

(S_2) the inferences supported by this background knowledge are secure to the degree to which the links between the present record (data) and antecedent causes or past events (background assumptions) are unique or deterministic, and to the degree to which the argument chains are relatively short and simple.

Once the evidence is *secure*, it must also exhibit epistemic independence in two senses:

(EI$_1$) background assumptions used to establish the evidence must be vertically independent of the hypothesis being tested; and

(EI$_2$) background assumptions (linking principles or sets of linking principles) derived from one or more different sources used to establish the evidential import of archaeological data must be horizontally independent of one another, i.e. no one set of linking principles entails the others as a proper subset of itself or is confirmed by the same evidence.

If the evidence is secure and independent, then archaeologists can triangulate, setting up a system of mutual constraint among lines of evidence bearing on a hypothesis – and ultimately bearing on the theory from which the hypothesis is derived. (If the hypothesis is confirmed, it will in turn help to confirm the theory.) So, if the two or more sources converge or fail to converge in support of one interpretation of the data, i.e. on one hypothesis that best explains the data, they together constrain the interpretation that is possible.

We can schematically diagram these conditions as follows:

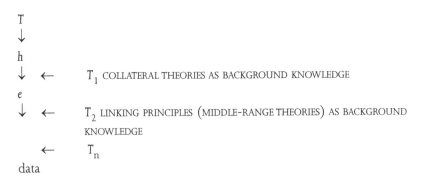

where the data are the material record; e is the interpretation of the data; T, T$_1$, T$_2$, and T$_n$ do not entail one another and are not confirmed by the same evidence; T$_1$, T$_2$, and T$_n$ are credible and to some degree uniquely determining of e and of h; and the evidence, i.e. the interpretation of the data, is secure and independent (S$_1$, S$_2$, EI$_1$, and EI$_2$ obtain).

Let us look at an example (for which a diagram would be very complex). Wylie's Consilience Model of Confirmation shows us that, in practice, researchers do not use arguments from only one subject-specific

theory, T, to test its hypotheses; instead, they often use background knowledge or "collateral" theories. For example, Wylie explains that New
Archaeologists hold an ecosystem theory (T) according to which human
behaviors – which leave the record archaeologists uncover – are system-
wide adaptive responses to the material conditions of life (Wylie 2002:
181). Faced with an archaeological record indicating that pueblos
throughout the US Southwest were abandoned by around 1300 CE as the
population aggregated in certain limited areas, New Archaeologists
hypothesized that the ecology of the Southwest had collapsed around
1300 CE. Some hypothesized that the collapse was not sudden and catas-
trophic, but was a result of a change in the rainfall pattern that led to
population decline and to social changes including pooling of regional
resources and eventually to aggregating in the places where agriculture
could sustain the remaining population. To find out, Hill and Longacre
developed a research program at two pueblo sites occupied between
1100–1280 CE.

From T, they deduced a materialist ecosystem hypothesis (h): the
dramatic aggregation and population decline were a culturally mediated
response to gradual but significant changes in the environment. Slow environmental changes set off local and restricted adjustments by people in the
Southwest with the cumulative effect of the large-scale change of pueblo
culture documented in the archaeological record. And from h, they
deduced several test hypotheses, i.e. hypotheses to test h and, ultimately, T:

(h_1) fine-grained shifts in patterns of resource exploitation were caused by
environmental pressure,
(h_2) variability of the artifacts at a site was caused by local change in social
structure and level of integration of prehistoric pueblo communities,
and
(h_3) regional trade networks were set up indicating a system of redistribution buffering people in areas of food shortage.

To test (h_1), researchers turned to paleoenvironmental science (T_1) and
found evidence of a region-wide shift in the pattern, not the annual
amount, of rainfall. Previously, the area experienced gentle, dispersed
winter rain, but this changed to torrential summer storms, causing more
erosion and less effective surface moisture. And turning to paleobotany
(T_2), they found evidence that maize production became restricted to

marginal upland areas after 1100 CE; in general, plant and animal (paleo-biology) remains showed a continuous decline in the use of wild plants and small game. They concluded that this decline plus population pressure resulted in local food shortages.

Hill and Longacre were especially concerned with (h_2). To show that aggregation of the population was a response to this environmental pressure, and not due to changes in dominant social norms, Hill argued that (h_{2a}), by 1100 CE, a matrilocal residence system and a matrilineal system of descent typical of later pueblos had already been established, but social integration had not yet reached the level typical of later pueblos; by 1100 CE, formerly autonomous and dispersed lineage units already co-existed in single-village settlements but kept their social distinctness. To support (h_{2a}), Hill interpreted ceramic data in the way we discussed above: (e) stylistic differences in the designs painted on ceramics found in three separate sectors of a pueblo reflected three different matrilocal groups of people.

To establish linking arguments between material and social variables – in our example, to make this last argument interpreting the ceramic data as (e), evidence for (h_{2a}), and so for (h_2) and so for (h) and so for (T) – archaeologists resort to analogies between ethnography and their data. Ethnographic studies (T_3) show that, in modern pueblos, women are the primary producers of ceramics and design styles are passed from mothers to daughters. By analogy (h_{2b}), our archaeologists argued that (h_{2a}) the local ceramic designs found at prehistoric pueblos were due to kinswomen living in matrilocal, stable residential groups. And our archaeologists also corroborated (h_2) the hypothesis that these pueblos were amalgams of village and homestead groups previously dispersed throughout the region and now co-existing near a stable water supply, by finding (e_1) patterns of stylistic difference and distribution in other kinds of artifacts typically associated with women's activities (Wylie 2002: 182).

This example shows us that, because archaeological data are inconclusive in themselves, archaeologists cannot deduce them from h alone (except when h is exclusively about biophysical conditions causing the record). So they use analogical arguments, not just deductive ones. But this is acceptable because they use different lines of evidence all of which constrain each other, and many of these lines are not from T, but from external theories, such as T_1 (paleoecology), T_2 (paleobiology), and T_3 (pueblo ethnography), as well as theories T_4–T_n about cultural evolution.

Thus, ecosystem theory T does not use just itself to establish and evaluate tests of itself. Ecosystem theory is incomplete, and, even if it were complete, Wylie notes that it could not specify relationships between all the variables archaeologists use to make their arguments. Independent bodies of theory must be used to reconstruct non-cultural variables such as environmental conditions and material constraints on resource exploitation, absolute dating, prehistoric technology, subsist practices, etc. But, in our case, these all converge on Hill and Longacre's hypothesis, and it is implausible that this consilience is an artifact of theoretical expectation, i.e. due to circular arguments.

Wylie points out that Hill and Longacre use tentative, *ad hoc*, substantive considerations to assess the credibility of h and the significance of *e*. There are no general laws specifying relations between sociocultural variables and material variables in the archaeological record, so they could not use structural considerations as Glymour's model predicts (Wylie 2002: 182). They use analogical arguments from substantive empirical information about *e* to show how it might have been produced. The structure of T and the structure of their inferences are, thus, unavoidably parasitic on substantive considerations. For example, Longacre uses substantive ethnographic data on pueblo ceramic production in his analogical argument from the archaeological data to his hypothesis about the social organization of prehistoric pueblos. He hypothesized that stylistic similarity in smallest units of design correlates with the intensity of social interaction; this is a linking hypothesis used to argue that pot shards found at the site correlate with different kinship units living at that site. And it is also used by Hill to interpret data regarding other kinds of artifacts at a second site as evidence of different kinship units living at the second site.

2.5 Gender considerations in archaeological reasoning

Wylie's model allows us to see how gender considerations, androcentric, sexist, or feminist assumptions can enter archaeological reasoning as researchers work to understand past cultures. Here we will examine three of the cases Wylie uses to show how this happens.

The work of Patty Jo Watson and M.C. Kennedy (1991) reveals pervasive androcentrism in explanations of the emergence of agriculture in the Eastern Woodlands of what is now the US (cited in Wylie 2002: 192). All of the main explanations of the shift from foraging to agriculture assume that women were responsible for gathering plants and small game under earlier foraging adaptations, and for cultivating domesticated plants once horticulture had been established. Nevertheless, Watson and Kennedy found a "glaring absence" of hypotheses postulating women as responsible for the development of cultigens.

One of the foremost models, T, makes the blatantly *ad hoc* hypothesis that shamans, assumed to be male, began horticultural practices because they used certain plants for ritual purposes and this led to further horticultural practices that transformed people in the Eastern Woodlands from foragers to agrarians. On this model, "[i]n effect, women passively followed indigenous plants around when foraging, and passively tended them when the plants were (re)introduced as cultigens by men" (Wylie 2002: 192).

The major alternative model T_1 postulates co-evolution: horticulture was an adaptive response to plants transforming themselves. Without deliberate human intervention, plants changed in ways people could use. This plausible model suggests that human garbage might have introduced artificial selection pressures that generated the varieties of indigenous plants that became cultigens.

Watson and Kennedy question basic assumptions of both models: that shamans were men, that dabbling with plants for ritual purposes was more likely to produce the knowledge and changes necessary for horticulture than systematically using them for food, or that no humans were responsible and these plants effectively "domesticated themselves." Against T_1, Watson and Kennedy point out that, if people developed the plants they wanted, then it is more likely that these people were the women who were primarily responsible for gathering food. Both models T and T_1 assume (no doubt unconsciously) that women could not have been responsible for any major culture-transforming process; this androcentric assumption (a corollary of the assumption that men are always the agents of cultural change) is uncritically taken from popular culture and from traditional anthropology (Wylie 2002: 193).

We see that, in the two androcentric explanations, the data were said to be significant, and so became evidence e for reasons entirely determined by theoretical assumptions about the cultural subject. The assumptions are that (T_{h1}) women are primarily responsible for plants as food while men are not; (T_{h2}) this sexual division of labor is natural; and (T_{h3}) women are passive, not agents of cultural change. And the evidence e was being used to test interpretations and explanations h using these same assumptions. (Schematically, T defines h and T defines e, which tests h and so tests T.) Watson and Kennedy thus show that two standard models (T and T_1) of horticultural transition in which women are not part of the explanation include sexist assumptions (T_{h1-3}) about the nature and capabilities of women. And the same assumptions (T_{h1-3}) are used in interpretations of data (e) used, in turn, to evaluate these models. This guarantees that the data will be seen as evidence supporting the models.

These assumptions (T_{h1-3}) are more important than constitutive virtues such as coherence, elegance, plausibility, and explanatory power, since the models T and T_1 insist on keeping the assumptions, even though, as a result, they are less constitutively virtuous, i.e. less coherent, less plausible, and have less explanatory power.

In their criticism of T and T_1, Watson and Kennedy use non-archaeological, independent background knowledge, conceptual and empirical, to get an independent evaluation of the framing assumptions of T and T_1. Conceptually, they find a straightforward contradiction in both models, viz. women are identified as tenders of plants, wild or cultivated, yet they are denied any role in the transition from foraging to horticulture. And Watson and Kennedy take empirical background knowledge from botany to show that the domesticated plant varieties used by people in the Eastern Woodlands routinely appeared in far from optimal prehistoric contexts. Therefore, against T_1, Watson and Kennedy argue that it is implausible to think that these domesticates arose in any way without human intervention. Too, they turn to the work of feminist anthropologists documenting the enormous variety of women's roles and the degree to which they are active, not passive; mobile, not home-bound; politically powerful, not dispossessed and victimized. This work shows the implausibility of the assumption (T_{h3}) that women are less capable of innovation and strategic manipulation of resources than males. And if this assumption is implausible, so are the interpretations (T and T_1) that depend on it.

This example also shows us where the old theory, T, succeeded and where it failed in making good inferences from the present to the past. As we have seen, on Wylie's Consilience Model of Confirmation, analogical inferences from present/source contexts to the past are secure when the material:behavioral link is credible in the present/source context and the link between surviving traces (data) and antecedent causes (h) is unique or deterministic. These requirements are met (and Watson and Kennedy agree with the other archaeologists) in making an analogy from the persistence of the association of women with horticulture across historically and ethnographically documented contexts to a past division of labor in which women are assumed to have had primary responsibility for agricultural activities. On the other hand, if there is evidence that the material:behavioral link is not stable in present/source contexts, then the analogy with the past is undermined and insecure. Model T (and, to an extent, T_1) failed to make a secure analogical inference when it used a background assumption not itself secured by the relevant background knowledge, i.e. the assumption (T_{h3}) that women are not agents of major change is not based on the relevant ethnography and ethnohistory which show women to be active, mobile, and resourceful.

A second case set out by Wylie provides an excellent example of the convergence upon one hypothesis of several horizontally independent lines of evidence. To show that gender roles and household structures are not the same everywhere at all times, but change as societal structures or dominant ideas change, C.A. Hastorf (1991) uses her work on pre-Hispanic sites in the central Andes (cited in Wylie 2002: 193). She argues that gendered divisions of labor and participation in the public, political life of these highland communities profoundly changed when the Inkas gained control of the region, well before the Spanish took over the region and began to form a state. Against other accounts, Hastorf argues that household structure and gender roles changed even before the Spanish came in and recorded them; thus, the Spanish did not encounter stable, traditional ways of life.

Hastorf compared the density and distribution of paleobotanical remains from households in the periods before and after the Inka gained control of the area. She found evidence that, over time, the female-associated activities of producing and processing maize increased and became more and more restricted to specific locations. She also found evidence that, before the Inkas came in, males and females had the same diet over their lifetimes. (Stable-isotope analysis of bone composition yields this evidence of variability in dietary intake.) And as evidence of Inka presence begins to appear, so does evidence that the diets of men and women began to differ: males ate more food with the isotope values identified with maize. Hastorf interprets the data using ethnohistoric records documenting Inka practices of treating men as heads of households and communities. The Inkas drew the men into ritualized negotiations using maize beer (chicha) and required their labor away from villages, and paid them with maize and chicha. Hastorf concludes that, throughout this period, the new political structures of the Inka empire forced a shift of gender roles on local communities and households: women produced more beer at specific locations and their participation in the society was more restricted.

The strongest security archaeologists can have when arguing for their hypotheses is provided when they can use completely independent, non-ethnographic sources that specify unique causes for bits of the record. We see an example of this in Hastorf's analysis of bone composition. Assuming background knowledge used in stable isotope analysis (T_1), she established the dietary intake necessary to produce the composition of the bone marrow in skeletons recovered by archaeologists. She also used paleo-botanical analysis (T_2) to show what plants and animals the people ate, and skeletal analysis to show sex-differences in what and how much they ate. Isotope analysis and paleobotany are independent of and so good tests of interpretations and explanatory presuppositions about subsistence patterns and hypotheses about gendered social practices affecting the distribution of food (Wylie 2002: 197).

Each line of argument from background theories T_1 and T_2 independent of the theory T under consideration (that gender roles and household structure changed with the advent of Inka control) is vertically secure. But no one of them alone is enough to make theory T credible. Thus, Hastorf uses more than one line of argument; she uses collateral lines of evidence T_1 and T_2 to establish (h) that the anomalous shift in diet in male skeletons was due to drinking more maize beer and to establish that this change was linked to (T) the advent of Inka-imposed systems of political control and changes in gender relations in the household.

E.M. Brumfiel (1991) found results similar to Hastorf's in the Valley of Mexico at the time when the Aztec state was forming a tribute system in the region (cited in Wylie 2002: 193). Brumfiel argues (T) that the Aztec state depended on tribute to maintain its political and economic hege-mony, and this depended on changes in the organization and deployment of predominantly female domestic labor.

Based on ethnohistoric and documentary evidence, she argues (h) that fabric production was largely the responsibility of women. And based on an analysis of the density and distribution of spindle whorls (used to spin yarn for weaving), she concluded (h_1) that fabric production increased dramatically in outlying areas and decreased in urban centers. And based upon this hypothesis she argued that the Aztecs extracted tribute in cloth payments (part of T).

She also analyzed the density and distribution of griddles and pots, and found that the proportion of griddles to pots changed over this period. More and more griddles were being used near urban centers (e); fewer were being used in outlying areas (e_1). On the other hand, she found a decreasing proportion of griddles in outlying areas (e_2); there pots came to predominate (e_3). Since griddles are associated with tortillas, which are labor-intensive but transportable, while pots are associated with less labor-intensive foods, Brumfiel interpreted the changing proportion of griddles to pots as evidence $(e-e_3)$ that (h_2) outliers came to prefer less labor-intensive foods and people closer to city centers produced transportable foods. On her interpretation, all this $(h-h_2)$ is evidence for her theory (T) that the primary burden of meeting tribute demands for cloth was shouldered by women and this caused shifts in their household labor. Cloth was extracted directly as tribute from the hinterlands and women took up less labor-intensive food preparation as they increased cloth production. People closer to a city center produced transportable foods to participate in markets there, including marketing food to a mobile labor force. Thus, (T) the organization of predominantly female domestic labor changed as the Aztec state came to depend on tribute to maintain its political and economic hegemony (Wylie 2002, 193–4).

As we have seen, the security of analogical inferences depends on (1) the security of the source-side analogue, i.e. the claim that the documented behavior of people has or had certain material effects; and (2) that these people are relevantly similar to those in the past. But the security of analogical inference also requires that the analogy fit the subject context in the sense that it adequately accounts for the archaeological record. Thus, Brumfiel's model and the alternative model both make an analogy from the association of women with spindle whorls used in weaving and griddles and pots used in food preparation in historically related ethnographic and ethnohistoric contexts to their association with women in prehistoric contexts, e.g. under Aztec rule. This analogy meets requirements 1 and 2. It is a stabilizing analogy inasmuch as the association between women and spindle whorls, pots, and griddles is agreed to in both models.

However, Brumfiel brings up not new data, but old data (viz., the changing distribution and density of spindle whorls and the changing proportion in the distribution of griddles and pots, e–e_3) that the alternative model cannot account for. In this way, she shows that the alternative model is not empirically adequate as an account of the political economy of the states that were formed in pre-Colonial Mesoamerica and South America.

She also reveals the implausible assumptions made in the alternative model, e.g. that gender structures are stable and unchanging (sometimes captured by saying they are "natural" or "traditional"). This is one of the assumptions that makes the alternative model inadequate. The other faulty assumptions are that political and economic processes of state formation are due only to male activities. Brumfiel's model tests the assumption that gender relations and household divisions of labor are dynamic, not natural, and that they are crucial co-determinants of political and economic processes of state formation.

Like Hastorf, Brumfiel uses horizontally independent lines of evidence: (h_1) she identifies the anomalous (by other models) distribution of artifacts related to cloth production over time and space, and (h_2) she reassesses evidence related to different sorts of food processing (the distribution and density of pots and griddles). Their unexpected convergence gives her theory strong support because nothing in the two sources of background knowledge and nothing about the hypotheses based on them insures that they would converge on her theory. The evidence could have turned out otherwise.

2.6 To be continued

Although naturalized feminist empiricist epistemologies are often distinguished from standpoint epistemologies, feminist philosophers of each persuasion have appreciated the insights, while remaining critical, of the other. Thus, it is not unusual to find aspects of standpoint theory in the work of feminist empiricists. However, Wylie's variant of standpoint epistemology takes up central tenets of standpoint theory, including one that has been problematic for feminist empiricism, viz. that some knowers have epistemic privilege. The reader needs some familiarity with classic standpoint epistemology of science to grasp Wylie's version and to appreciate the problems her version solves for feminist empiricists; therefore, we will present Wylie's feminist empiricist standpoint epistemology of science in our discussion of standpoint epistemologies of science found in Chapter 5.

3

FEMINIST PHILOSOPHY OF SCIENCE
AND FEMINIST VALUE THEORY

How sharp is the dichotomy between facts and values? Clearly they function in different ways in our thinking and in our lives, but how differently? Logical empiricists made a very sharp distinction based on the conviction that factual claims can be corrected by experience and values cannot. The sciences offer the most rigorous, methodical ways to verify and correct factual claims and accounts of our world, but since values cannot be falsified or verified, they lie outside the sciences. They must, therefore, be held "dogmatically" as expressions of opinion or emotion.

Holding to this picture, it also seems that, if moral or political values are taken as considerations in a scientific research project, they must compete with evidential and other cognitive considerations for control of the inquiry. That is, scientists use *either* facts *or* values to guide research; but not both. At best, contextual values (moral, social, or political values and interests) displace attention to evidence and valid reasoning; at worst, they lead scientists to bias, wishful thinking, dogmatism, dishonesty, and totalitarianism (Anderson 1995b: 33–4). In these ways, contextual values are supposed to interfere with the goal of scientific work, which is the discovery of truth. There are no logical or evidentiary relations between value judgments and factual judgments. Thus, since contextual values have nothing to do with whether a hypothesis or theory is true, they cannot serve as criteria for choosing among competing hypotheses or theories. If they do serve as criteria for theory choice, the result is bad science. Anderson reconstructs a logical argument for this traditional model of the unhappy relationship between evidence and contextual values that captures many of the assumptions lying behind it:

1   Significant truth is the sole aim of theoretical inquiry.

2   Whether a theory is justified depends only on features indicative of its truth, not its significance.

3   One shows that a theory is (most probably) true by showing that it is (best) supported by the evidence.

4   A theoretical proposition is supported by the evidence only if there is some valid inference from the evidence (in conjunction with background information) to it.

> (H: A theoretical proposition is supported by the evidence only if there is a good fit between the evidence, the background assumptions, and the proposition.)

5   Value judgments take the form "P ought to be the case."

> (H: And value judgments are like wishes and desires.)

6   There is no valid inference from "P ought to be the case" to "P is the case" (or any other factual truths).

7   There is no valid inference from value judgments to factual truths (5, 6).

> (H: There is no good fit between value judgments, background assumptions, and factual truths.)

8   Value judgments can provide no evidential support for theories (4, 7).

9   Value judgments can play no role in indicating the truth of theories (3, 8).

10   Value judgments can play no role in justifying theories (1, 2, 9).

(Anderson 1995b: 33–4)

Philosophers accept different formulations of this argument. Anderson points out that Haack probably does not accept (4). Instead, Haack holds 4(H), the relation between evidence and theory is one of "fitness;" data support a claim to truth if there is a good fit between the data and the claim. Moreover, she understands values to be like wishes or desires. Thus, Haack's assumptions (H) are included in Anderson's argument.

Haack and others express the point that values are not evidence for theories by endorsing premises (5) and (6). But Anderson notes that focusing on (6) is a red herring because no defender of value-laden inquiry ever suggested that it is legitimate to infer "P" from "P ought to be the case." Feminists certainly do not think that because the Equal Rights Amendment ought to be the law of the land it is so! "The real contests," Anderson says, are over premiss (1) the claim that the goal of all theorizing is truth, and premiss (2) the claim that whether a theory is justified depends only on features indicative of its truth, not its significance. Against these claims, Anderson argues that the aims of theoretical inquiry are "broader than the bare accumulation of truths, and the justification of theories is relative to all these aims," i.e. merely collecting facts does not add up to good science. Scientists need facts that are relevant to the purposes or aims of their investigations. *The justification of theories depends on these broader aims as well as on other truth-conducive features.* This leaves "an opening for moral, social, and political values to enter into theory choice" (Anderson 1995b: 37).

3.1 Feminist philosophy of science and feminist value theory

Haack and other philosophers recognize that the aim of research is not just to find a lot of true statements, not just "the bare accumulation of truths." No one is interested in a jumble of facts. The interests of the researchers (often interacting with the interests of their funders) determine the questions to which the research is addressed. In medicine, for example, interest in most research questions is driven by the positive moral and social value we place on human health. The aim of a piece of research is to answer a question like "Does this gene correlate significantly with breast cancer?" and, ultimately, "How can breast cancer be cured or eliminated?" The research aims to find not every single fact about this gene and every single fact about breast cancer, but those facts that are relevant to the question asked. The facts must be *significant*.

When a research report does not give us the significant facts, it is likely to give us a partial or biased answer to the research question, even if each sentence in the report is true. Thus, research should give us not just a theory of the phenomena, but an *adequate* theory of the phenomena, on pain of bias and partiality. Anderson takes as an example the book *The Secret Relationship between Blacks and Jews*, which purports to uncover the role of Jews in the Atlantic slave trade. She notes that what counts as a significant truth is any truth that bears on the answer to the research question, in this example, "What was the role of Jews in the Atlantic slave trade?" The "whole truth" is "all the truths that bear on the answer" or "a representative enough sample of such truths that the addition of the rest would not make the answer turn out differently." And when research questions are motivated by contextual interests or values, in this case, moral values, then what counts as significant truth and whole truth can only be judged in relation to these interests (Anderson 1995b: 39–40).

The implicit question driving the argument of *The Secret Relationship* is, "Do Jews deserve special moral opprobrium or blame for their roles in the Atlantic slave system or bear special moral responsibility for that system's operations?" To answer this question, an adequate theory would give us all the facts morally relevant to it, "or enough of them that adding the rest would not change the answer." But *The Secret Relationship* ignores many morally relevant facts, e.g. those showing that "Jews behaved no differently, from a moral point of view, than anyone else who had the opportunity to profit from the slave system." Therefore, the book offers a biased account.

Haack and others hold that significance and impartiality can be defined solely in relation to truth, for example "an unbiased account of the Jews' roles in the Atlantic slave system is simply one that truthfully represents their roles in their 'actual' proportions relative to other ethnic groups." Thus the driving question would be, "How did Jewish roles in the slave system compare with the roles of other ethnic groups?" But this question does not specify which roles and comparisons are of interest. "Is it more important that a greater proportion of US Southern Jews owned slaves or that they owned fewer slaves per capita [than other groups]?" Note well, we are deciding which proportion to use as a *factual* criterion of significance or impartiality. But this decision itself depends on the question motivating the inquiry. In this case, the question is laden with contextual values. Thus, deciding which comparison is significant has *moral* import, from which it follows that *the decision as to which factual proportion is significant is not and cannot be value-neutral*. We must engage in substantive normative theorizing to determine the bearing of facts upon the normative claim in question.

Anderson's analysis of the problem with *The Secret Relationship* differs from Haack's. Haack would say the problem is letting moral and political values guide research, particularly the selection of significant facts. Anderson says that there are two problems; one is the moral and social values serving as the background for framing the research question and selecting significant facts; that is, we can morally assess these values and they may turn out to be pernicious. The second problem is that the book is not impartial; it "doesn't count as inquiry, because it is rigged to reach a foregone conclusion." The research fails to live up to the pragmatic requirement that the answer to the research question be genuinely open to the evidence and arguments, both those that favor the investigator's preferred conclusion and those that undermine it (Anderson 1995b: 41).

The ideal of "value-neutrality" would leave research without any direction because, without values and valued interests, researchers cannot distinguish a significant from an insignificant fact and a biased account from one that is impartial. Thus, the proper ideal is impartiality, not value-neutrality. Impartiality is not achieved by disregarding evaluative standards; instead, impartiality is achieved by "a commitment to pass judgment in relation to a set of evaluative standards that transcends the competing interests of those who advocate rival answers to a question." We have seen that *a significant fact is one that bears on the answer to the research question and an adequate answer to the research question is comprised of all the significant facts or a representative sample of them such that the addition of the rest would not change the answer.* Moreover, when the research question is motivated by contextual interests or values such as moral or political values, i.e. by evaluative concerns, then *a significant fact is one that has evaluative bearing.* Thus, impartiality in such a case requires judgment that meets evaluative standards and these *evaluative standards* include *honesty and fairness in judgment,* where fairness "demands attention to all the facts and arguments that support or undermine each side's value judgments" (Anderson 1995b: 42).

3.2 A co-operative model of theory justification

To show how contextual values can legitimately influence theory choice, Anderson (1995b) sets out a co-operative model of the justification of theories. It models the interaction between normative and evidential considerations in theory choice, revealing the use of contextual values in answering evaluative research questions and the use of evidence in evaluating contextual values. The model applies to any research in the natural or social sciences directed to evaluative questions.

1   All scientific work begins with a question. Questions direct research by
    (i) defining what counts as a significant fact, viz. one that bears on the answer to the question
    (ii) defining what counts as an adequate account of a phenomenon, viz. one that captures enough of the phenomena in question that the addition of further detail won't change the answer.

Contextually value-laden questions yield theories subject to three criticisms:

(A) All the statements of T are true, but T is trivial, insignificant, does not address the interests motivating the question.

(B) All the statements of T are true but T is biased, incomplete, or distorted, viz. it pays disproportionate attention to pieces of significant evidence that incline toward one answer, ignoring significant facts supporting rival answers. If the question has moral import, whether T is biased depends on the moral relevance of the evidence cited. This assessment depends on moral and political value judgments.

(C) T may be addressed to a question that has illegitimate normative presuppositions.

2 When the questions are based on contextual interests, answers must address the interests, thus these contextual values direct research by

(i) shaping the description/classification of research objects, viz. grouping together phenomena that bear a common relation to these interests. This is because purely epistemic criteria of significance are not sufficient to define theoretical classifications.

Theories using such classifications are subject to the epistemic requirements that

(E_1) there are clear empirical criteria for deciding when phenomena fall under a classification,

(E_2) some phenomena do fall under the classification, and

(E_3) the classification must figure in some explanation or causal or empirical regularity.

Theories using such classifications are subject to the normative requirements that

(N_1) classifications track the underlying contextual values accurately, i.e. group phenomena together that share a common relation to these values or interests. (T may misconceive relevant, legitimate interests and classify together phenomena that should be separated or exclude phenomena that should be included in a class.)

(N_2) the contextual values themselves are ethically justified. (T's classifications may be based on illegitimate contextual values and should be rejected altogether, e.g. sexist and androcentric values.)

(ii) shaping the methods used to answer them, i.e. methods must be adequate to reveal the phenomena deemed significant by those interests.

Theories using such methods are subject to the criticism that they use methods that foreclose the possibility of discovering valuable potentialities or important differences or similarities among the objects of inquiry.

This model pictures the *co-operative* relationship between the normative and evidential considerations underlying theory choice inasmuch as

(a) contextual values set standards of significance and adequacy (and, thus, of impartiality, lack of bias) for a theory, and evidence determines whether the theory meets the standards;
(b) contextual values help define what counts as a meaningful classification and the empirical criteria for identifying things falling under it, and evidence determines what if anything meets these criteria; and
(c) contextual values help determine the methods needed to answer a question, and the evidence gathered using those methods helps answer it (Anderson 1995b: 53–4).

In Haack's view, truth is the only goal of science that is relevant to justifying theories because only evidence is truth-conducive. It therefore competes with contextual values, social interests, etc. that are not truth-conducive. In Anderson's view, there are many goals of scientific inquiry other than finding truth. Some of these goals are based on contextual values because "modern science exists in large part to serve human interests." The point of scientific theories is to organize hypotheses through models that select important from unimportant phenomena. And what is deemed important "depends on practical needs and interests which may be gendered or staked in other socially constructed positions such as class or race" (Anderson 1995b: 53; 1995a: 56; see also Tiles 1987).

In this way, factual criteria of significance and impartiality are justified in relation to the background interests motivating a research question. When these are moral or political interests, the criteria need moral and political justification. And since these factual criteria set legitimate standards for theory choice, "moral and political values legitimately figure in the justification of theories" (Anderson 1995b: 42).

3.3 The scope of the co-operative model

Isn't Anderson's model limited to the social sciences in which research questions are driven by social interests? Philosophers have offered accounts of significance as determined by internal factors, not external, moral, or political ones. Anderson takes up Philip Kitcher's suggestion that, in the natural sciences, significant statements are those that answer significant questions, i.e. those that "challenge the basic explanatory schemata of a theory – either to show that the schemata can be widely and effectively instantiated or that their presuppositions are true" (Anderson 1995b: 43; Kitcher 1993: 95, 112–13). Anderson's response is to point out that,

> The constitutive goals of many natural sciences include the promotion of particular contextual values. The constitutive aim of medicine is the promotion of health; of horticulture, the advancement of our abilities to grow food and other useful plants; of engineering, the construction and manipulation of useful artifacts. We rightly judge the significance of questions and answers in these fields in relation to these practical interests.
>
> (Anderson 1995b: 43)

This is one reason why many philosophers prefer to use physics as the paradigm of value-neutral science. But Anderson is skeptical of the claim that physics is a "pure" science since political and social interests so clearly drive much of the research in physics, e.g. into the questions, "under what conditions will a mass of fissionable material enter into an uncontrolled nuclear reaction?" And even questions in mathematics such as "what algorithms can rapidly factor very large numbers?" are "significant only because states and businesses have political and commercial interests in constructing and decoding encrypted messages."

Anderson concludes that there is "no clear way to isolate a special subset of sciences or fields of inquiry in which no such interests play a role in defining significance." Feminists do not need to argue that all science is value-laden. But given that very good physics, mathematics, medicine, horticulture, and other sciences are influenced by values, it seems silly, as Anderson points out, to keep up the search for a model of good science as value-neutral, i.e. neutral among contextual values (Anderson 1995b: 43).

3.4 Values, interests, and natural kinds

The question whether contextual values influence scientific work in the context of justification is tied up with some deep philosophical issues. Nowhere do we see this more clearly than in attempts to limit contextual values to the "context of discovery." The context of discovery is the way in which a proposed hypothesis is discovered – which may include contextual values – but is supposed to be irrelevant to the "context of justification," i.e. the way in which a proposed hypothesis is proven using good methods of observation, data selection, analysis, testing, etc. and which is assumed to be free of contextual/non-cognitive values, though it may legitimately include constitutive values/cognitive virtues. Thus, the fact that moral and social interests in human health drive us to ask research questions such as, "Does this gene correlate significantly with breast cancer?" arises in the context of discovery. But as long as the context of justification is immune to the influence of moral and social values, scientific research can be conducted so as to achieve its aims, viz. the discovery of nature's structures, its natural kinds, and their objective dependencies, e.g. causal relations and statistical correlations. What a gene is, what a breast is, what cells are cancerous, etc. are all questions about what kinds of things exist in nature. This view that there are natural kinds is sometimes expressed by the claim that nature has "joints" and that good science "cuts nature at the joints."

The best arguments for the possibility of classifying natural kinds hold that, if science groups phenomena into classes according to whether phenomena in one group are causally connected to phenomena in another group, it will track natural kinds, cut nature at the joints (Kitcher 1993: 71). Since causal regularities in nature exist independently of our interests and values, natural kinds do, too. It follows that if we try to use our interests to discover natural kinds, we might coincidentally discover some, but we would miss most of them; however, if we use causal regularities as criteria of epistemic significance, we will cut nature at the joints and track natural kinds.

In this view, we can use our interests to classify kinds of things, e.g. medicine groups organisms into pathogenic and non-pathogenic because we are interested in our health. But there is a true classification, perhaps captured by the evolutionary concept of reproductive fitness (e.g. pathogens reduce human reproductive fitness and so have a common causal impact on our evolution), or some other as yet undiscovered concepts (proponents of this view hope that science asymptotically approaches the truth in such matters). The sciences should find such criteria of epistemic significance since these criteria reveal nature's own classifications.

Perhaps this view is correct about high-energy particle physics. But Anderson holds that the world is too complex and messy to be organized into a few all-inclusive mutually exclusive classes accounting for all causal regularities. She argues (with John Dupré 1993 and others) that scientific classifications are partly justified by contextual interests. For each theoretical classification based on a standard of epistemic significance, e.g. its members have a genuine causal relation to some other phenomenon, it is likely that there are some other cross-cutting classifications in the neighborhood that bear a causal relation to yet other phenomena. So criteria of epistemic significance alone do not tell us which classifications to base our theory on. Because other classifications in the neighborhood could equally satisfy this standard of epistemic significance, some further justification is needed for theorizing the world in terms of the classification selected. And the justification may well come from contextual values (Anderson 1995b: 44–6).

3.5 Contextual values and good science

Defenders of the value-neutrality of science also have a psychological argument to show that contextual values are bad for science: if a scientist uses values, e.g. feminist values, in her research, she will not be impartial when she assesses evidence about the hypothesis. Anderson quotes Max Weber, who said, "Whenever the person of science introduces his personal value judgment, a full understanding of the facts *ceases*" (Weber 1946: 146, cited in Anderson 2004: 4). At best, the psychological mechanism of wishful thinking will make her misunderstand the facts (Geertz 1990: 19, cited in Anderson 2004: 5). At worst, she will be dishonest or close-minded and reject an impartially justified hypothesis because it "inconveniences" her "political program" (Gross and Levitt 1994: 162; cf. also Haack 1993: 37–8, both cited in Anderson 2004: 6 and 8).

In Anderson's analysis, these psychological arguments for the value-neutrality of science depend on a contradictory and crude model of how value judgments work. Haack, Gross and Levitt, and others hold the contradictory views that (*a*) values are science-free, i.e. facts cannot be evidence for or against values and yet they hold that (-*a*) feminists dismiss facts *that count against their values*, which clearly assumes that facts can be evidence against (and so, for) values. They also hold the crude view that value judgments are inherently dogmatic, that people hold to their values regardless of the facts. But the claim that "values are dogmatic" does not fit well into the dichotomous view of facts and values. It is not a true factual judgment because we see people take their experiences as evidence for and against value judgments. And as a value (or a value recommendation), "thou shalt hold values dogmatically," the claim that value judgments are inherently dogmatic is absurd (Anderson 2004: 22). We will return to Anderson's response to this view below.

Of course, defenders of the value-neutrality of science do recognize that contextual values affect science in ways that do not, as they think, compromise its rationality and its discovery of truth – though they are certainly not truth-conducive. Anderson sets out five such ways:

1   in the context of discovery;
2   in the context of justification when necessary, e.g. humane treatment of experimental subjects;
3   in the context of use, e.g. interests and values may determine the use of results;
4   in practical reasoning to determine the best means to our ends; and
5   assessing how far some values are realized in fact (Anderson 2004: 6–7).

Anderson argues to the contrary that if contextual values are properly theorized, we see how *they can be truth-conducive*.

3.6 Using experience and facts to correct values

Feminist philosophers have argued that when two competing theories are both empirically adequate, scientists can and sometimes do use contextual values to decide between them (Longino 1990; Potter 1988 and 2001). In these arguments, feminists have assumed that contextual values, whether feminist or sexist, etc. are "on a par epistemically." This was sufficient (1) to show that theory choice could be based on both empirical adequacy and contextual values thereby producing good scientific theories, and (2) to make room for feminist science, i.e. choices among empirically adequate theories based on feminist values. But Anderson points out that feminists have not been able to distinguish among values based on their empirical fruitfulness. We need to distinguish legitimate from illegitimate uses of such values, whether feminist or non-feminist. She does not think all moral and political values have equal epistemic value, nor is it the case that a contextual value that is legitimately used and fruitful in one inquiry is always legitimately used and/or fruitful in all research. It follows that feminist values are not always legitimately used in science, and that patriarchal values are not always illegitimately used. In this way, value judgments in science have been undertheorized.

Constitutive values such as empirical adequacy, simplicity, scope, consistency with other accepted theories, etc. have legitimate epistemic uses in scientific work, viz. they are used along with evidence to justify theory choice. Anderson argues that contextual values can also have epistemic uses in scientific work. To aid her argument, she considers, with Hugh Lacey (1999: 2–6), two senses of the thesis that science is value-free:

1   A scientific theory (T) is *neutral* if it does not
    (a) presuppose any contextual intrinsic value judgments, or
    (b) support any contextual intrinsic value judgments.
    (A value judgment is intrinsic if it presents something as valuable "in itself" or "for its own sake;" a value judgment is extrinsic if it presents something as valuable "for the sake of something else." Thus, the satisfaction of basic human needs might be an intrinsic value and helping others in time of need might be an extrinsic value, valuable ultimately for the sake of satisfying basic human needs – which is valuable in itself (Zimmerman 2003).)
2   T is *impartial* if the reasons for accepting T are impartial among rival contextual values. Thus, the only reasons for accepting T are T's relations to the evidence and T's having some constitutive value(s) such as empirical adequacy, scope, simplicity, or conserving established theories. Here the claim for the value-freedom of science is that there are grounds, viz. evidence and constitutive values, for anyone to accept T, regardless of their contextual values.

(Anderson notes that it is very hard to formulate the claim of impartiality, but assumes for the purpose of the argument that it could be suitably qualified and soundly formulated.)

We should note that T can be impartial without being neutral. In particular, T can be adopted impartially and yet not be (a) presupposition-neutral, i.e. T might presuppose value V. Anderson points out that it seems that if T presupposes V, then V must be the reason for accepting T, but someone might hold V, yet accept T, not because of V, but because of the evidence and/or constitutive virtues of T. Although cognitive virtues are (supposed to be) rationally binding, contextual values are not (Anderson 2004: 4 n2)

Anderson's model reveals that good science can presuppose, can be guided by, contextual values despite the commonly held view that 1(a), i.e. good scientific theories must be neutral, not presupposing contextual values. T might presuppose V in a way that, for example, leads the researchers to classify their data according to V; here V provides a norm for classifying data. If T is more epistemically fruitful than rival theories that refuse to classify their data according to this norm, then V makes T more epistemically fruitful. So using values in science is sometimes epistemically justified. That is, scientists accept T on impartial grounds, because it is more fruitful than rival theories, not in spite of T's presupposing a contextual value, but because of presupposing it! How can this be so?

Defenders of the value-neutrality of science often hold that it is the job of our practical reason to figure out the best means to our ends. Someone's "final end" is whatever that person takes to be valuable in itself or intrinsically valuable. Such ends are determined by our desires etc., which are unaffected by our (theoretical or pure) reason, and since intrinsic values are determined by our desires etc. and not by empirical evidence or reasoning, values are science-free. This is roughly captured by saying, "You cannot derive 'ought' from 'is.'" Nothing could ever count as evidence that some things are good or bad. And this is why so many philosophers take the view that values are inherently dogmatic, that we all hold our values dogmatically and cannot do otherwise (Anderson 2004: 6 and 8).

Anderson argues to the contrary that growing up, having human experiences such as disillusionment, etc. allows most people to learn from experience that some of their values are mistaken. Most people are capable of growing and learning in these ways. Some people are not; these people are dogmatic, holding to their values regardless of the facts, or, perhaps, just holding to some values regardless of facts. But values themselves are not inherently dogmatic. One of the primary reasons that most people can learn from experience that their values are mistaken is because we take our emotional experiences – which Anderson defines as "affectively colored experiences with people or things or events" – to provide evidence that these people or things or events have value; for example, if we experience California redwoods with awe, we take this as evidence that they are splendid (Anderson 2004: 9). While philosophers admit that we do this, traditionally they have been skeptical of whether emotional experiences are reliable sources of evidence for the value of anything. Many have assumed that they could never be.

We cannot treat Anderson's rich arguments in detail here, but she agrees that some emotional experiences have cognitive content (Deigh 1994); that is, some experiences are "affectively colored experiences," and, like most experiences, these have cognitive, usually representative, content. Moreover, such experiences are defeasible (though not as responsive to the world as beliefs), i.e. we can find out that the representative content is erroneous, confused, etc. Thus, if we find out that the cognitive content of an emotional experience is defective in some way, we might discount the importance of the feeling, too.

Such emotional experiences can function as evidence for values because these experiences are independent of our desires and ends. In Anderson's example, Diane desires elected office and values a political life. Despite her desires and values, she feels badly about the political life, disillusioned by campaign financing, political backbiting, and small political gains. These emotional experiences do not depend on her desires and values, and, in fact, undermine them. But an ally tries to persuade her that her "disappointment with what seems to be a merely symbolic victory reflects an unduly narrow perspective." Taken in isolation this victory achieves little, but "in the long view it can be seen as fundamentally shifting the terms of debate. What seems like a hollow victory is a watershed event. This [factual] judgment could be tested over a longer stretch of experience." The ally argues that Diane should continue to value the political life. This sort of persuasive argument is quite common and makes sense only because our emotions are responsive to facts. *And usually our emotions are reliable, though certainly not infallible, evidence for our value judgments.* (The exceptions include emotions affected by drugs, depression, etc.) When it is clear that the representational content of an emotional experience is adequate, we can trust our emotions. "Indeed, we would be *crazy* not to." And if we do not trust our emotions, if we hold to our values despite the facts and despite our feelings, we *are dogmatic* (Anderson 2004: 9–10). We may conclude, then, that *values are not "science-free" and so need not be held dogmatically.*

3.7 Legitimate and illegitimate uses of contextual values in science

If values are not "science-free" and need not be held dogmatically, then values might legitimately be used in science. But how shall we distinguish legitimate from illegitimate uses of values in science? Anderson suggests as a criterion that values are legitimately used in science if *the values do not drive research to a predetermined or favored conclusion*. In her model, the influence of facts and values is bidirectional, and when empirical research in social (or natural) science is addressed to evaluative questions, "especially about the relations of various phenomena to well-being" (e.g. does divorce help or hurt the people involved?), evaluative presuppositions do not determine the answer to the evaluative question; rather, the evidence determines the answer. The value presuppositions help uncover evidence relevant to the question. How do they do so?

To get at the legitimate use of values, Anderson distinguishes three sorts of bias caused by the illegitimate influence of values on science: a research design is

(B_1) "biased in relation to the object of inquiry if it (truthfully) reveals only some of its aspects, leaving us ignorant of others." Anderson notes that this sort of bias is inevitable because all research designs close off some lines of research, but this bias is harmless as long as we do not think the research has covered all aspects of the object of inquiry.

(B_2) biased in relation to its hypotheses if it is rigged (wittingly or not) to confirm them. A good research design must allow its hypothesis to be disconfirmed by evidence. Although proponents of value-neutral science claim that all value-laden research will only confirm researchers' evaluative presuppositions, Anderson rightly points out that it is not the values guiding the research that cause bias in relation to its hypothesis; instead, it is the failure to use proper methods, the precautions regularly taken in research such as drawing fair samples of evidence or treating controversial results symmetrically, i.e. not stopping research when one makes findings that support one's hypothesis, but putting the hypothesis through further tests.

In this respect, Anderson is arguing that research guided by feminist values is just good science as ordinarily practiced: "From an epistemological and methodological point of view, research guided by evaluative presupposi-tions functions just like research guided by any other presuppositions." But she departs from the traditional model of value-free science in showing that value-laden research can still be good science. And she leaves it open whether research into evaluative questions can ever be guided by value-free presuppositions.

(B₃) biased in relation to a controversy if it is more likely to (truthfully) uncover evidence that supports one side rather than all sides. On the other hand, one "research design is more fruitful than another, with respect to a controversy, if it is more likely to uncover evidence supporting (or undermining) all, or a wider range of sides of the controversy" (Anderson 2004: 18–20). (It remains an open ques-tion, then, when a research design is just less fruitful than others and when it is biased in relation to a controversy.)

Thus, *contextual values legitimately influence science if*

1 precautions are taken to avoid these three biases, and
2 the values are *epistemically fruitful*, i.e. they guide research "toward discov-ering a wider range of evidence that could potentially support any (or more) sides of a controversy." Thus, a contextual value is more epistem-ically fruitful than others if it has more power to uncover *significant* phenomena. When a less fruitful value guides research, important evidence can still be uncovered, but we must remember that such research is limited to answering only certain questions or giving only a partial answer to a controversial question (Anderson 2004: 20).

The remaining issue, of course, is whether the research is epistemically fruitful *because* of the contextual value(s). The (presupposition) value-neutrality thesis, 1(a), assumes that all the epistemic work is done by factual elements; therefore, if the research is epistemically fruitful, it cannot be on account of contextual value(s). Anderson argues that, assuming we can distinguish the factual and normative components of a thick evaluative judgment, i.e. distinguish the empirical features of the world it picks out from its claim to normative authority, then we can ask whether the epis-temic fruitfulness of a thick evaluative judgment is due to its normative authority. (A thick evaluative judgment is one that simultaneously expresses both factual and value judgments, e.g. "S is rude" both describes S's behavior and evaluates it negatively.)

Anderson's model makes use of a hortatory division of the stages of research and in the model we see that *evaluative presuppositions, evidence, and evaluative conclusions interact at each stage of research.*

Researchers

(a) begin with an orientation to the background interests animating the field;

(b) frame a question informed by those interests;

(c) articulate a conception of the object of inquiry;

(d) decide what types of data to collect;

(e) establish and carry out data sampling or data generation procedures;

(f) analyze their data in accordance with chosen techniques;

(g) decide when to stop analyzing their data; and

(h) draw conclusions from their analyses.

(Anderson 2004: 11)

3.8 A case study

Let us look briefly at the example Anderson uses to illustrate her model of research as epistemically fruitful precisely because it is guided by contextual values. For her case study, Anderson turns to feminist research on divorce carried out by Stewart *et al.* (1997). Stewart *et al.* did a longitudinal study of consequences of divorce for the people involved. The primary reason for research on divorce is to discover evidence that can help us evaluate practical recommendations regarding divorce. That is, we seek evidence that will inform our practical recommendations regarding divorce by informing our value judgments about various factual aspects of it. But, as we have seen, we gather many of these facts because our research is guided at every step by one or more contextual values.

(a) The background interest animating this field of research is uncovering the effects of divorce on the well-being of those involved, i.e. the spouses and children, if any. Here Anderson compares two orientations towards this background interest: one is feminist and one she dubs the "traditional family values" model of the family. In the traditional model, husband and wife are married for life, live in the same household, and raise their biological children. The roles of parent and spouse are inseparable: the wife is the mother of her husband's children and vice versa. This is held to be best for the children and probably for the parents. On this orientation, divorce "breaks up" the family and harms the children.

Anderson found that feminists are ambivalent about divorce; sometimes it seems to enable men to leave their wives and so disadvantages the wives, and sometimes it seems to allow women to escape oppressive marriages. Too, feminists question whether the "traditional" family should be the norm for post-divorce relationships among those involved.

(b) "The different value orientations of traditionalists and feminists suggest different research questions" (Anderson 2004: 12). The traditionalist question is, "Does divorce have negative effects on children and their parents?" To answer this question, researchers are likely to compare members of divorced and non-divorced families on measures of well-being, especially negative ones, e.g. sickness, poverty, and behavior problems. Anderson also found that traditionalists tend to favor methodologies that focus on aggregate differences among research subjects and to do main effects analyses. Main effects analyses aim to uncover the main effects of independent variables (e.g. the divorce) on the outcome (e.g. measures of well-being).

The feminist researchers note that it is virtually impossible to distinguish the effects of divorce from the effects of marital problems leading to divorce (Anderson 2004: 12–13). Even divorced and non-divorced families with similar problems differ in other important respects. Moreover, the feminists thought that (1) focusing on negative outcomes makes it hard to find positive outcomes; (2) focusing on aggregate differences between divorced and non-divorced groups assumes that the findings, and evaluations based on them, apply to each person in the group; and (3) focusing on divorce as an event assumes that it has a fixed, enduring meaning, and misses whether its meaning changes over time – positively or negatively, but especially positively, e.g. a divorce may "recede in significance as individuals cope with it and engage the new experiences that it makes possible" (Stewart et al. 1997: 30, quoted in Anderson 2004: 13). Thus, the feminists asked, "How do individuals differ from one another and over time regarding the effects of divorce on them and the meanings they ascribe to it?"

(c) Their value orientations also function as background assumptions influencing the way researchers conceive of the object of inquiry. Thus, Wallerstein, a clinical psychologist, argues that divorce scars children for life (Wallerstein and Kelly 1980; Wallerstein et al. 2000). Her "conception of divorce drawn from a clinical perspective focuses on the individual's problems with an event in the past, stressing its negative aspects" (Anderson 2004: 13). Here divorce is conceived in terms of "trauma," "loss," etc. These are thick evaluative concepts, expressing both factual and value judgments, e.g. a "trauma" is a sudden injurious event; to name something an "injury" is both to describe a fact and to evaluate it negatively – which is non-controversial when the injury is physical, but socially and even morally normative when applied to psychological states. In the same way, a "loss" is a good thing that is now lacking. Anderson notes that, although such a conception of divorce is evaluative, it is nevertheless legitimate (it does not rig the results) and fruitful – it guides researchers to look for certain kinds of evidence using appropriate methods, e.g. measures of psychological disturbance (Anderson 2004: 14).

The feminist conception of divorce is complex, differing from the "traditional" conception in many ways. Here we mention only two. (1) The feminist team saw divorce as both loss and "opportunity for personal growth," and not as a traumatic, sudden event, but as "an extended process of adjustment to a new set of life circumstances that could go better or worse over time" (Anderson 2004: 14). Thus, the feminist perspective required a longitudinal study – which was legitimate because it did not rig the results by guaranteeing that researchers would find change over time. (2) The feminist team "conceived of divorce not as breaking up the family, but as transforming it by separating parental from spousal roles." The post-divorce family can then be seen as a family with co-parents living in different households. This enabled the researchers to compare models of the post-divorce family to see which are better for children.

(d) Their value orientations also influence what data researchers will gather. Some standard measures of well-being relevant to the value of divorce include financial security, children's behavioral problems, physical illness, etc. "Traditionalists" gather data using these measures. Feminists include, in addition to these traditional measures, individuals' feelings about their situation and their emotionally colored interpretations of their situations as evidence of the individuals' well-being and the value of divorce. The feminist background assumption here is that individuals have privileged, though not infallible, normative authority to assess their own well-being. Thus, the feminist team gathered data including subjects' self-assessments, e.g. many divorced women who were left with lower incomes were nevertheless pleased to have greater decision-making power over their money.

(e) Data sampling can also be influenced by background values. Evaluative conceptions of divorce as a loss or an opportunity for growth, as breaking up the family or transforming the family, should not guide research in ways that simply confirm themselves. One way to avoid the self-confirmation of guiding values is through careful sampling procedures. Good sampling methods require that fair samples be drawn, e.g. the dependent variable should not be the basis for selecting one's sample. Unfortunately, when Wallerstein drew her sample from among people who attended a psychological treatment clinic, she drew a sample that was biased toward people who were having problems with divorce. By drawing its sample from divorce dockets, the feminist team drew a less biased (though not perfect) sample.

(f) Researchers must decide which among all their variables are significant and which relationships among the variables are significant, and, since they cannot analyze every relationship, they have to decide which ones to analyze and how to analyze them. Background values influence such decisions, for example whether to use a main effects analysis – which focuses on the main effects of independent variables on the outcome, and whether to measure aggregate differences between groups, thereby discounting individual variation and taking the average outcome to represent all members of the group; or to look for interaction effects – focusing on the interaction among independent variables and the effects of the interaction on the outcome. Thus, researchers who hold that there is a single best way of life for everyone, such as the "traditional" family life, will employ a main effects analysis, suggesting that their findings represent all members of families. But researchers who hold that different ways of life, e.g. different forms of "family," are better for different people, will want to pay attention to "within-group heterogeneity" and so will employ an interaction effects analysis.

(g) When should researchers stop analyzing their data? It is tempting to keep analyzing unwelcome data and to stop when the results are pleasing. But good science requires a symmetrical approach to data analysis: researchers should treat welcome and unwelcome results symmetrically, especially when the results are controversial. Thus, using a main effects analysis, the feminist team made the welcome finding that "divorced mothers were better adjusted if they worked full-time." But because they decided to treat their welcome findings the same way they treated unwelcome ones, they analyzed their data for interaction effects and found that mothers "who were working prior to the divorce did much better if they continued working after the divorce. But mothers who had previously stayed at home did worse if they went to work after the divorce" (Anderson 2004: 17; Stewart *et al.* 1997: 100–1).

(h) The conclusions of divorce research make normative assessments, e.g. positive and/or negative evaluations of divorce and of the effects of divorce, and so can be used to make normative suggestions, e.g. about how best to cope with divorce and its effects. The research is carried out to answer evaluative questions on the basis of empirical evidence and, as Anderson points out, would be senseless if science could not support values, i.e. if "ethics were science-free."

Is the epistemic virtue of a contextual value ever due to its normative authority? Proponents of (presupposition) value-neutrality say "no." But Anderson presents a case of research in which evaluative presuppositions, e.g. individuals who have privileged – not infallible – normative authority to assess their own well-being, were used to guide researchers to collect and analyze data that included individuals' self-assessments. And it is the normative validity of this value presupposition that directly explains its epistemic value. "It is precisely because subjective emotional responses and emotion-laden interpretations are normatively relevant to judgments of well-being that [including such subjective measures made the feminists'] research more fruitful than research programs that focus only on objective measures" such as finances and behavior problems (Anderson 2004: 21). Using traditional measures uncovered interesting facts about the negative outcomes of divorce on women and children, but using subjects' feelings and interpretations uncovered, for example, the positive outcome that 70 percent of the women judged their personalities to have improved since divorce (Anderson 2004: 15; Stewart et al. 1997: 66). This significant finding could be made only because of the feminists' value presupposition.

3.9 Conclusion

Anderson's model shows that our answer to evaluative questions will not be *empirically* adequate, i.e. describe phenomena adequately, without using *normatively* adequate evaluative presuppositions to guide research. One consequence of the model, shown clearly in the case study she presents, is that researchers can presuppose feminist values (among others) and still be impartial, be open to evidence against their favored/valued outcomes, and so to whatever answers they discover.

4

FEMINIST CONTEXTUAL EMPIRICISM

Helen Longino understands a good epistemology as both setting out the conditions that must be satisfied in order for people to ascribe the status "knowledge" to something and describing what people do. A good philosophy of science, then, describes the work of scientists and gives a detailed theory of the conditions that must be satisfied in order for the scientific community to *know* something rather than *hypothesize* it (Longino 2002: 10). The claim that philosophers should describe what people do, e.g. their knowledge-producing practices, is contested by philosophers for many reasons: many wish to maintain a strong distinction between philosophy as "conceptual" and social or natural science as "empirical," and/or maintain a sharp distinction between facts and values, and/or believe that descriptive epistemology and philosophy of science leads to pernicious relativism. But Longino belongs to the school of epistemologists and philosophers of science who wish to *naturalize epistemology and philosophy of science* inasmuch as she believes that philosophy must give accounts of knowledge in general and scientific knowledge in particular that are based more closely on what scientists (or other knowers) actually do when they produce and transmit knowledge. But Longino does not belong with those naturalizers who think that philosophy should no longer concern itself with norms (i.e. the conditions that must be satisfied in order for people to ascribe the status "knowledge" to something) and should not produce normative accounts of knowledge. Philosophy should be both normative and empirical. This is why she sets up *The Fate of Knowledge* as breaking down the dichotomy between social and rational accounts of knowledge (Longino 2002: 10).

Holding a sharp distinction between facts and values, it seems to many philosophers that, unless science is value-free, it will not establish the facts (assuming that the purpose of science is to establish facts); instead, values, functioning as biases, will lead to bad science. To show that values and interests can and often do operate in scientific work that is accepted and well-respected by other scientists, Longino argues that scientific knowledge is produced through practices carried out primarily by communities of scientists, that background assumptions are always at work in scientific reasoning from evidence to hypotheses, models, and theories, and that values and interests – including assumptions about gender – can and often do function as background assumptions. For example, the positive valuation of sex/gender dimorphism, that there are and should be two sex/genders, functions as a background assumption and appears in knowledge produced by natural and social scientists.

4.1 Evidential relations

Contra the claims of positivists and others, Longino argues that there is no unique or intrinsic evidential relationship between evidence and the hypothesis or model for which it functions as evidence. Instead, the connections or regularities we appeal to in assessing evidential relations are connections or regularities from some point of view and are always subject to change. Thus, people take *e* as evidence for a hypothesis *h* when they believe *e* belongs to a class of things that are related (causally or through class membership) to the class of things *h* belongs to. The objects, events, and states of affairs providing evidence for hypotheses do not carry labels showing what they are evidence for. Instead, how one determines evidential relevance depends upon one's background beliefs or assumptions (Longino 1990: 43 and 45).

Longino describes a background belief as "an enabling condition of the reasoning process." It enables us to see *e* as evidence for a hypothesis *h*, not in the sense that we come to believe *h* *simpliciter* (although we might) but in the sense that we come to believe that, given the background assumption *b*, *e* makes *h* plausible. As Longino puts it, background assumptions are "beliefs in the light of which one takes some *x* to be evidence for some *h* and to which one would appeal in defending the claim that *x* is evidence for *h*" (Longino 1990: 44).

4.2 Constitutive and contextual values

In keeping with other empiricist philosophies of science, Longino distinguishes two types of values. The first type, *constitutive values*, such as truth, accuracy, simplicity, predictability, and breadth, are generated from an understanding of the goals of science, and are the source of rules determining what constitutes acceptable scientific practice. *Contextual values*, on the other hand, are personal, social, and cultural values and interests, expressions of preference about what ought to be; these belong to the social and cultural environment in which science is done (Longino 1990: 4). Although contextual values are supposed not to affect good scientific work, Longino distinguishes five ways in which they *can* and she describes several cases in which they *do*.

Contextual values can

1   affect practices that bear on the integrity of science, for example when the desire for profit leads scientist/entrepreneurs to present results first at a press conference rather than in a professional journal or at a professional conference;
2   determine which questions are asked and which ignored, for example when scientists are led by concern about world-wide overpopulation to emphasize the therapeutic advantages of a contraceptive drug over its hazards;
3   affect the description of data, that is, value-laden terms may be employed in the description of experimental or observational data and may influence the selection of data or of kinds of phenomena to be investigated;
4   be expressed in or motivate the background assumptions facilitating inferences in specific areas of inquiry; or
5   be expressed in or motivate acceptance of global, framework-like assumptions that determine the character of research in an entire field.

On Longino's account, then, gender considerations enter the content of scientific theorizing when assumptions about gender are included among the background beliefs used to determine the relevance and strength of the evidence for a scientific hypothesis.

4.3 Case studies

The feminist uses to which contextual empiricism can be put are revealed in several cases Longino offers. Here we will briefly summarize one from evolutionary studies and give an extended presentation of another from behavioral endocrinology.

Recent studies of human evolution exhibit the third and fourth ways in which contextual values can enter scientific work, viz. in the description and selection of data as well as in the background assumptions facilitating inferences from data to hypotheses. Recently, evolutionary studies have produced two approaches to explaining human descent from primates. The androcentric, "man the hunter" perspective assigns a major role to the changing behavior of males. On the androcentric account, the development of tool use is understood as a consequence of the development of hunting by males. The gynecentric, "woman the gatherer" perspective assigns a major role to the changing behavior of females; in particular, this approach argues that females developed tool use as a response to the greater nutritional stress experienced by females as abundant forests were replaced by less abundant grasslands and by changing conditions of reproduction (Longino 1990: 107–8).

On the androcentric account, fossil bones, teeth, and objects identified as tools function as evidence for hypotheses about men developing stone tools and spears, and developing smaller canines, larger brains, co-operative behavior, and language. On the gynecentric account, these bones, teeth, and stones function as evidence for hypotheses about women developing gathering implements, stone tools for softening hard fibers or crushing seed pods, and developing larger brains, co-operative behavior, and language. To facilitate the inferences from fossil teeth and bits of stone to human behavior and evolution, anthropologists make analogies with contemporary hunters and gatherers. But depending on which contemporary groups one chooses, one gets very different pictures of human evolution. As Longino says,

> Man-the-hunter theorists will describe the role of the chipped stones in the killing and preparation of other animals, using as their model the behavior of contemporary hunting peoples. Woman-the-gatherer theorists will describe their role in the preparation of edible vegetation obtained while gathering, relying, for their part, on the model of gathering behavior among hunter/gatherers.
>
> (Longino 1990: 109)

Each approach provides an example of a contextually driven background assumption facilitating inferences from data to hypotheses. How the data are read depends on whether one is working within the framework of man the hunter or woman the gatherer. These background assumptions provide the basis for assigning relevance to the data. The fossil bones, teeth, and stones alone do not tell us anything about the behavior of our evolutionary ancestors. On the contrary, as Longino states,

> If female gathering behavior is taken to be the crucial behavioral adaptation, the stones are evidence that women began to develop stone tools in addition to the organic tools already in use for gathering and preparing edible vegetation. If male hunting behavior is taken to be the crucial adaptation, then the stones are evidence of male invention of tools for use in the hunting and preparation of animals.
>
> (Longino 1990: 111)

A second case study is based on the extensive literature in behavioral endocrinology published between the early 1970s and mid-1980s. Longino (1990, Chapters 6–8) treats all three categories of sex differences thought by researchers to be correlated with and caused by sex hormones: (1) anatomical and physiological, (2) temperamental and behavioral, and (3) cognitive; however, limitations of space permit us to discuss only the first two categories.

(I) Data relevant to anatomical and physiological differences between the two sexes in humans come from (i) the correlation of male and female bodies with higher and lower than average levels of the sex hormones – androgens and estrogens – in the body and (ii) the correlation of abnormal, sex-linked anatomical and physiological characteristics (e.g. hermaphroditism) with very high or very low levels of sex-hormones. There are also (iii) animal studies of the effects of varying hormone levels in fetuses and in live animals (Longino 1990: 113).

Hypotheses about the development of anatomical and physiological differences between male and female human reproductive systems are fairly well established. Human fetuses have the potential to develop into either males or females until, during the third and fourth month, exposure to an androgen brings about the development of the internal and then external organs of the male reproductive system, and lack of such exposure brings about the development of female internal and external sex organs (Longino 1990: 115). (The importance of estrogens has become apparent through more recent research.)

Hypotheses about the development of sex differences in the central nervous system and the brain are not yet as well established. Examples of such hypotheses are H_1 "androgen receptors play a primary role in sexual differentiation of the human brain," and H_2 "peripheral gonadal hormones alter the sensitivity to neurotransmitters such as serotonin."

Summarizing inferences from data to hypotheses about hormonal determination of (1) anatomical sexual differences, Longino says,

> The human studies do support the idea that exposure of the primordial tissues to testosterone or one of its metabolites, for example, 5-α dihydrotestosterone, at the appropriate time is both necessary and sufficient for masculine development of the sex organs. This inference is further corroborated by experimental data in a variety of mammalian species whose reproductive anatomy and physiology are analogous to those of humans. In contrast with the animal experiments used to support claims about behavior, the systems here really are analogous: penises, testicles, seminal vesicles are all quite similar across mammalian species. Furthermore, the anatomical effects of hormone exposure or its failure in the animals studied are invariable rather than probabilistic.
>
> Researchers have also been able to establish in large part the biochemical pathways of action of testosterone. The similarities in the physiological systems involved and in the relation of the presence or absence of testosterone to sex organ development allow the model of action established in nonhuman mammalian species to be applied to humans as well. The uncertainties here have to do not with the functional interactions of different anatomical areas but with completion of the biochemical pathway analysis. For example, because the exact mechanism of hormonal action at the cellular level is only partially under-stood, it is not yet certain how testosterone or its metabolites act in the cell nucleus. . . . The lack of certainty will be allayed by more information and more analysis. The biochemical data we do have, however, make the hypothesis regarding male sex organ development that we have been discussing as unassailable as biochemical theory itself. This, of course, is not to say that biochemical theory is unassailable but that the assumptions

involved belong to a well established body of theory currently in use.

(Longino 1990: 126–7)

However, in the inferences from data to hypotheses about the anatomical/physiological effects of gonadal hormones, one of the most important background assumptions facilitating the reasoning is the *assumption of sex/gender dimorphism*, i.e. the assumption that there are two sex/genders. In studies of hormonal influence on the differentiation of external genitalia in humans, "[t]he current view is that testosterone secreted by the fetal testis is required for normal male sex organ development and that female differentiation is independent of fetal gonadal hormone secretion." As we have seen, the most important human observations are of persons affected by various hormonal abnormalities. But even here we find that the *assumption of dimorphism* – that there are and should be two sexes – affects the description of the data:

> Cases diverging from prototypical male or prototypical female development are treated as cases of partial or incomplete male or female development. Such individuals are treated as inadequate males or females rather than as instances of types with their own integrity or as points on a continuum of which prototypical males and females are the extremes. . . . Genetic males who lack intracellular androgen receptors and are thus unable to utilize testosterone exhibit a female pattern of development of external (though not internal) genitalia. Genetic females exposed in utero to excess androgen . . . exhibit what is described as partial masculine development, including enlargement of the clitoris and incomplete fusion of the labia.
>
> (Longino 1990: 126)

Turning to effects of hormones on brain development and so on later behavior, we also find the *assumption of dimorphism*. Researchers including Ehrhardt *et al.* refer to the "masculinizing," "demasculinizing," "feminizing," and "defeminizing" effects of gonadal hormones on brain development to make an analogy with the effect of these hormones on the development of different reproductive systems in males and females. But Longino points out that these terms are used inconsistently to refer to

(1) behavior, (2) "hypothesized but unknown determinants of behavior," and (3) "specific events in neural development." These terms have fairly clear meaning when applied to behavior, viz. they refer to behaviors stereotypically attributed to men and women in particular cultures. But when these terms are applied to processes of neural development – hypothesized *but not known* to exist, researchers are led to select

> some aspects of the biochemical processes they can trace at the expense of others. For instance, that the gonadal hormone involved in brain development in males starts out as testosterone is made more important than the fact that it is aromatized to estradiol [*an estrogen*] before being taken up by neurons.
>
> (Longino 1990: 122–3)

Richard Whelan suggests that grouping gonadal hormones into androgens and estrogens reflects *hypotheses* about how they work even though scientists *often do not know* how they work. The background assumption here is that there are two very different sexes:

> brain organization research seems governed by a conviction of the deep differences between the sexes involving sharp biochemical distinctions and cleanly separable lines of masculine and feminine development. This conviction has resulted in a classification system for gonadal hormones generated by their purported effects rather than by chemical structure or mode of action. . . . The nomenclature both masks the complexity of hormone action and leads people to think of gonadal hormones as themselves male or female.
>
> (Quoted in Longino 1990: 123)

(II) Establishing the effects of sex hormones on brain organization is central to most hypotheses about the effects of sex hormones on behavior because the former, if sound, provides a central mechanism for the latter, "a mechanism that mediates between the genotype and its behavioral expression" (Longino 1990: 118).

Longino shows us that the best data relevant to claims that hormones determine behavior come from (1) experiments on animals to determine the effects of hormone levels on reproductive behaviors such as mounting, and on other behaviors such as fighting behavior in laboratory animals. There are also (2) (somewhat controversial) studies of social and behavioral differences between human males and females, and (3) (more reliable) clinical studies of people with hormonal irregularities, e.g. "young women with CAH (congenital adrenocortical hyperplasia), a condition leading to the excess production of androgens during fetal development," and other groups. The data from these groups support hypotheses about the *organizing* role of fetal hormones on later behavior, i.e. the hormones affect the development of the individual, irreversibly programming tissues "to respond in certain set ways to later physiological events." Finally, there are (4) studies of the *activating* role of sex hormones in *humans*, i.e. the hormones are circulating in the mature individual and trigger certain effects in the way that adrenalin increases the heartbeat. These studies include problematic attempts to correlate testosterone levels with levels of hostility and aggression (for example, in male prisoners). Studies of the activating role of hormones in *animals* have been somewhat more successful, but there are questions about inferences from animals to humans, as we shall see (Longino 1990: 112–14).

Researchers hypothesize that sex hormones affect behavior in three areas: sexual orientation, gender identity, and gender role behavior. For example, "attempts have been made to attribute homosexuality to both prenatal and circulating endocrine imbalances: deficiencies in the sex-appropriate hormone or excess amounts of the sex-inappropriate hormone." But here we will follow Longino's focus on research into effects of sex hormones on gender role behavior, e.g. energy expenditure, playing at parenting, social aggression, career choice, etc. conducted by well-respected endocrinologist Anke Erhardt and her collaborators (Longino 1990: 115).

Erhardt attributes gender role behavior to prenatal hormone exposure arguing from both animal studies and from observations of girls with CAH. Longino explains,

> CAH involves the excess production of androgens by the adrenal gland. It also involves a failure by the adrenals to produce cortisone. In fact girls with CAH are born with large clitorises sometimes mistaken for penises and usually surgically altered in later life. All individuals with CAH require lifelong cortisone treatment to compensate for their nonfunctioning adrenals. The

majority of the CAH girls studied are described as exhibiting "tomboyism," characterized as a behavioral syndrome involving preference for active outdoor play (over less active indoor play), greater preference for male over female playmates, greater interest in a public career than in domestic housewifery, less interest in small infants, and less play rehearsal of motherhood roles than that exhibited by "normal" young females.

These descriptions of the girls' behavior may have been influenced by "observer expectations" since they were obtained from the girls themselves and from parents and teachers who knew about the girls' condition, but Longino assumes "for purposes of this analysis that the reports are more or less accurate" (Longino 1990: 118–19).

Erhardt hypothesizes that these girls' gender inappropriate behavior is caused by the higher levels of androgens to which they were exposed prenatally, and more generally that "behavior thought appropriate to one gender (or sex) but not to the other, is importantly influenced by prenatal exposure to sex hormones." Her inference from correlation to causation (from the correlation between prenatal exposure and later sex role behavior to the claim that the hormones cause the behavior) is supported (1) by citing work intended to show that analogous behaviors in other mammals are caused by sex hormones, and (2) by agreeing with the hypothesis that sex hormones influence brain organization and suggesting that this is the mechanism through which prenatal exposure to hormones gets expressed in (i.e. causally influences) behavior.

(1) The analogous behaviors include, for example, more mounting behavior by female rats exposed prenatally to higher levels of androgens than unexposed females, and less mounting behavior by castrated male rats than by uncastrated males, as well as more (or less) fighting behavior in laboratory cages by rats prenatally exposed (or not) to higher levels of androgens. But, as Longino notes, it is unclear that the animal and human behaviors are analogous, e.g. that fighting behavior by animals in cages is analogous to rough and tumble play and outdoor play by children. It is the *background assumption of sexual dimorphism* in behavior, i.e. that there is female behavior (e.g. playing mommy) and male behavior (e.g. rough and tumble play), that does the work and makes the analogies "obvious" (Longino 1990: 120).

(2) Though Erhardt *et al.* acknowledge that the evidence for sex hormones influencing the development of the brain and central nervous system in humans is "tentative" and that "the role of social learning is much greater in human behavior than in subhuman mammals," they maintain that "there is sufficient evidence to suggest that biological factors influence psychosexual differentiation in human beings, too" (quoted in Longino 1990: 121). The basic hypothesis here is that "[b]rains are sexually differentiated anatomically and thereby predisposed to produce sexually differentiated behavioral responses." Longino mentions two such hypotheses: H_1 exposure to hormones during a critical period "affects the threshold of response within neurons to environmental and hormonal stimuli in later life," and H_2 "hormones affect the development of the neural circuits responsible for certain behaviors." We have seen that the assumption of sexual dimorphism facilitates inferences in research on brain organization (Longino 1990: 122).

We also find this assumption that there are two sex/genders in the reasoning of scientists from data to hypotheses about the behavioral effects of gonadal hormones. Longino suggests that masculinity and femininity are "ideals of personhood," i.e. models

> to which we aspire or whose realization is urged upon us. . . . Such ideals characterize individuals more or less imperfectly. To attribute the status of ideal to a description implies that the better a person exemplifies that description, the better sort of person she or he is. The content of an ideal may be identical with that of a stereotype.

But whereas stereotypes are meant to describe, ideals are prescriptions. Thus, we stereotype males and females when we *describe* them "as though all men equally exemplified masculinity and all women femininity." But these concepts clearly function as *ideals* when we derogate "those who depart noticeably from them" as, for example, "sissies," "tomboys," "bitches," or "wimps." These people are thought to be either moral failures or victims of nature. *The assumption of gender dimorphism, then, is both a fact and a value* (Longino 1990: 167–8).

Longino points to cultures with non-dimorphic gender categories, e.g. some past Native American cultures for which "berdache" was a distinct category to make the point that gender dimorphism is not inevitable (Longino 1990: 169). And biologist Anne Fausto-Sterling suggests that there are at least five sexes:

> medical investigators have recognized the concept of the intersexual body. But the standard medical literature uses the term *intersex* as a catch-all for three major subgroups with some mixture of male and female characteristics: the so-called true hermaphrodites ... who possess one testis and one ovary (the sperm- and egg-producing vessels, or gonads); the male pseudohermaphrodites ... who have testes and some aspects of the female genitalia, but no ovaries; and the female pseudohermaphrodites ... who have ovaries and some aspects of the male genitalia but lack testes. Each of those categories is in itself complex; the percentage of male and female characteristics, for instance, can vary enormously among members of the same subgroups. . . .
>
> Recent advances in physiology and surgical technology now enable physicians to catch most intersexuals at the moment of birth. Almost at once such infants are entered into a program of hormonal and surgical managements so that they can slip quietly into society as "normal" heterosexual males or females.
>
> (Fausto-Sterling 2004: 477)

Arguably then, gender dimorphism is not a fact, but it is clearly highly valued, for, like homosexuality, hermaphroditism is understood to be a pathology. Like most people, researchers assume that there are "two types of individual: one with female reproductive capacity, feminine behavior, and a sexuality oriented towards men, the other with male reproductive capacity, masculine behavior, and a sexuality oriented towards women." Moreover,

> as long as dimorphism remains at the center of discourse, other patterns of difference remain hidden both as possibility and as reality. In particular, the idea that there could be a multiplicity of modes of personality organization linked to sex and sexuality – a multiplicity of genders constructed at the intersections of biological sex, sexual orientation, reproductive status, class, race, and sexual ideology or morality, for instance – remains submerged.
>
> (Longino 1990: 170 and 171)

4.4 Relativism

Many philosophers reject the view that a good philosophy of science should be empirically adequate to the history and current practices of science because making the epistemology of science empirically adequate seems to lead to relativism. If a *philosophy* of science recognizes the empirical fact that different scientific communities produce knowledge in different ways and even produce conflicting measurements, models, and theories, and if philosophers do not find criteria for deciding which of these is correct (or at least which is the best measurement, model, or theory), then we shall be forced to say that all the measurements, models, and theories are equally valid, or that a scientific community's findings are true for that community. Thus, creationism would be just as valid as the evolutionary synthesis; it would be true for the Christian community just as the evolutionary synthesis would be true for the secular humanist community.

To show why this radical relativism is not implied by her account of knowledge, Longino adds to traditional empiricist criteria for distinguishing knowledge from hypothesis or opinion and gives us a new way to talk about and evaluate the relationship between what is known and the way we express what is known. Along the way, she makes a number of distinctions among (1) senses of "knowledge"; (2) theories of who knows; (3) views about what is known; (4) kinds of relativism; and (5) senses of "social." Individual *radical* relativism is avoided by her social account of knowledge, but recognizing that the individuals who make up cognitive communities are socially located, and building social location into an epistemology requires some relativism because, as a matter of fact, different cognitive communities, including different scientific communities, have different views that bear on what the community decides it knows. Is this still too much relativism?

4.5 Epistemological individualism

Let us begin with who knows. Epistemological individualism holds that it is the individual who produces knowledge and the individual who knows; it also assumes that all individuals are epistemically the same. A group knows in the (secondary) sense that the individuals who comprise it all know (or most of them, anyway). In this view, the only way to avoid radical relativism, i.e. the view that an individual knows whatever he thinks he knows, is to find criteria of knowledge that distinguish it from opinion. Philosophers have been trying to do this since the beginning. And the history of twentieth-century philosophy of science is strewn with efforts to find the criteria distinguishing scientific knowledge from pretenders to that title. The problem has been that philosophers' criteria are abstract and work at a very general level, but, when it comes down to real historical cases, those cases we look upon as *good* scientific work very often do not fit the abstract philosophies of science produced by philosophers. This has led most sociologists of science and historians of science to conclude that there is no one philosophy of science that distinguishes real scientific knowledge from pseudo-science. This is not to say that scientists do not themselves make a distinction, but they don't make it according to an abstract logic of science. And this fact has led many important sociologists and historians of science to relativism. With some exceptions, most sociologists do not buy into individual relativism, the view that an individual knows whatever he thinks he knows and all individual knowledge is equally valid. Most recognize that science is produced by groups or communities of scientists who decide among themselves on what they think is true or, anyway, the best scientific account of whatever it is they work on. Thus, the community sometimes declares the work of an individual scientist to be unconvincing. But absent any universally accepted norms or rules for what counts as true scientific knowledge, many important science scholars have concluded that scientific knowledge is comprised of whatever a scientific community thinks is true. And since there are no norms or rules for adjudicating among competing accounts of the same phenomenon, all are equally epistemically warranted and equally valid.

Faced with the unpalatable fact that different scientific communities do sometimes arrive at conflicting measurements, models, and theories of the same phenomena, many philosophers of science continue to search for norms that will determine the superiority of one theory or set of theories (even though scientists may not yet have discovered this set). Longino agrees with these traditional philosophers that we should avoid radical relativism and find norms that distinguish scientific and other knowledge from opinion, but she differs from them in that she is not an epistemological individualist. In her view, knowledge is social; therefore, it is guided by social norms.

4.6 Monism and pluralism

Behind the traditional philosophers' search lie two assumptions: epistemo-logical individualism or a failure to take seriously the social nature of scientific (or, indeed, any) knowledge, and monism. Monism makes several assumptions: the aim of science is to produce a unified account of the natural world and the ultimate success of science consists in producing that account. These assumptions in turn rest on the assumptions that the models or theories scientists produce when they investigate a natural system fully capture (or aim to) the causal processes constituting that system and that all the models or theories that scientists produce can be joined together into a coherent set or, perhaps, into one great theory or model.

Monism can be contrasted with pluralism of the sort Longino endorses; pluralism makes different assumptions from those monism makes: the natural world is too complex to be captured through a single theoretical approach and pictured in one unified account and not all science aims at producing such an account; instead, science has many aims and goals, most of which are specific and local (Longino 2002: 93 and 142). Ultimately, Longino's best argument for pluralism is to revisit a number of important cases of good scientific work and show that these cases are best explained by a pluralist view, not by monism. This is an appeal to relatively unreconstructed historical cases and reflects her (naturalized) view that a philosophical account of the aims of science and the theories of science should itself be empirically adequate, fitting cases of good scientific work (see Longino 2002: 175–83).

Pluralism says there is no a priori requirement that different approaches to the same area yield compatible observations or measurements, or that only one approach can be correct. However, conflicting data need not indi-cate deep incommensurability between different theoretical approaches. Conflicting data may just be

> signs that different approaches are producing different bodies of knowledge of the same complex system, each of which conforms to that system differently, as both Mercator and Peterson projec-tions produce two-dimensional maps that conform, but differently, to the topography of the spherical planet Earth.
>
> (Longino 2002: 201)

The traditional search for criteria by which to determine the one true (or best) model or theory also assumes that (since science aims at truth) all the true statements/accurate models science produces must be consistent. But, as a contextualist, Longino holds that statements cannot be detached either from their truth conditions or from the context in which their truth conditions are determinable; for example, a

> measurement made in the context of atomic physics is carried out in relation to standards of precision and comparison specific to atomic physics. It can't be compared to a measurement made in the context of quantum chromodynamics or experimental high-energy physics without the construction of another context that provides criteria for the comparative assessment of measurements from the different approaches.
>
> (Longino 2002: 94)

We cannot compare measurements from atomic physics and measurements from high-energy physics without a third context that gives us the terms in which to compare them.

Longino does allow the possibility that, as a matter of fact, it might turn out at the end of inquiry that one, unified account of the world is the best/true one. But she reminds us that it might turn out that such unity and consistency cannot be found, either because the world is not simple enough or because humans are incapable of finding the unified account. Longino argues that the question whether the world is too complex to be captured in a single unified account (pluralism) or simple enough to be captured in one theory or a consistent set of theories (monism) is an empirical question and should not be decided a priori by an epistemology or philosophy of science. A good epistemology/philosophy of science should not foreclose the question of monism/non-monism without argument (Longino 2002: 95).

Finally, the search for criteria for the one true or best scientific account assumes epistemological individualism, for it does not take seriously the fact that scientific work is social and that it is the social nature of science that mitigates against radical relativism. By "social" Longino means that scientific knowledge is produced when individuals interact in the proper ways with others in their scientific community. We will explain this below.

4.7 Knowledge as social

Why should we believe that knowledge practices are social as opposed to individual? Our answer depends in part upon whether we are naturalists, holding that an epistemological analysis should be an analysis of actual scientific practices, or non-naturalists, holding that an epistemological analysis should be an idealized, abstract account and that actual scientific cases can be rationally reconstructed to fit our account. This, in turn, depends on the aims of epistemology/philosophy of science. One aim might be an epistemological analysis that is useful to people, perhaps so that they can use it to arrange their knowledge-producing practices so as to improve them and find out more about the world for the sheer joy of knowing or to maximize their ideals of a better life. If this is so, then the philosophical analysis cannot be so idealized or abstract as to be irrelevant to actual knowledge-producing practices. On the other hand, if the aim is to produce an abstract model for the sheer fun of finding one that covers as many rationally reconstructed cases of good knowledge-production as possible, regardless of its usefulness to anyone, then the analysis need not be relevant to actual practice. Roughly speaking, the greater the commitment to its usefulness, the more necessary it is that a philosophical analysis be empirically adequate to actual knowledge practices.

Longino and many other epistemologists accept the fact that scientific knowledge is socially produced, maintained, and transmitted. Their acceptance is based upon the reports of sociologists and historians of science, upon their own observations of scientists at work, and upon arguments, such as Wittgenstein's private language argument, showing that making a distinction between knowledge and belief, or true claims and false claims, or any number of other distinctions essential to having knowledge, or, indeed, to having any language at all, requires a community of at least two people. The fact that scientific knowledge is social, along with the conviction that epistemology and philosophy of science should be empirically adequate, i.e. naturalized in the sense mentioned above, leads Longino and others to set out an account of scientific knowledge as social.

Empirical studies of science show us that the production of knowledge is social inasmuch as scientists rarely work alone and even if one does, she cannot just declare that, when she is satisfied with her results, she has produced scientific knowledge. Until she has, for example, presented her results to members of the scientific community of which she is a part and has convinced them, so that they cite her work or use her results, she is not in fact deemed to have produced knowledge. There are many processes that scientific hypotheses and models must go through before they are accepted by scientists. Scientists "design and execute particular experiments on particular occasions for particular purposes, they count a particular set of specimens with particular measurement technologies, and they select particular sites for particular field studies." Along the way, they engage in such practices as "particular ways of manipulating experimental material – for example, transferring samples from vials to petri dishes, particular ways of measuring those samples, particular kinds of question about and particular categorizations of the samples, particular statistical methods" (Longino 2002: 98). Philosophers often lump these practices together as "observation."

Too, scientists

> interpret observations and experiments, they support or critique conjectures or hypotheses, they derive consequences, they extend models to new domains. They have multiple reasons for the particular choices and decisions they make in the course of all these activities, reasons that include feasibility, potential for application, aesthetic values, interest from other colleagues, interest from potential consumers, intelligibility to colleagues, resonance with metaphysical or ideological commitments.

Philosophers often lump these together as "reasoning" (Longino 2002: 98).

Empirical studies, as Longino points out, show that in the course of these activities, lumped together as observation and reasoning, scientists make many decisions for which they have reasons such as "feasibility, potential for application, aesthetic values, interest from other colleagues, interest from potential consumers, intelligibility to colleagues, resonance with metaphysical or ideological commitments." "These," she notes, "are the kinds of factors included under the umbrella of 'the social.'" Longino's view is that *all of these activities are social* in the sense that they are interactive, by which she means that they involve discursive interactions among different "voices"/people (Longino 2002: 98–9).

For example, sociologists of science point out that it is through critical discussion or social negotiation that scientists decide what the data are. Thus, Karin Knorr-Cetina and Klaus Amann (1990) describe how a typical group of scientists in a molecular biology lab take a fuzzy film and give a series of interpretations of it, until, finally, they decide that the bands are here and here, and not there. Then, they synthesize several such films (rather like cutting and pasting) until they produce one suitable for the publication of their results. As Longino remarks, "There is no guarantee that the individuals involved would eventually and independently have arrived at the vary same reading rather than one of the others compatible with the evidence of the film" (Longino 2002: 100). The film does not force the ultimate reading of itself.

Empirical cases, however, are not enough to convince philosophers that scientific observation and reasoning are social. Thus, Longino argues that, although an individual can make an experiment or record his perceptions alone, the status of the perceptions as *observation* or as *data*, depends upon their being stable enough that others similarly placed with similar equipment would see the same thing. Data must be accessible and invariant across the community of scientists. This is why experiments and observations must be repeatable by others (even though they may not be repeated) (Longino 2002: 101–2). The stability and reliability of data is assured by subjecting results to the criticism of peers, a paradigmatically social activity.

One major form of scientific reasoning (along with calculation and constructing new ideas) is using data as evidence to support theories and hypotheses. And, once again, the decision whether something is an appropriate reason is made socially, through discursive interactions, i.e. scientists in the relevant field critically discuss the matter among themselves, e.g. in publications, conference presentations, poster sessions, seminars, informal talks, telephone conversations, etc. These discursive interactions are all part of scientific challenge and response: a claim is put forward; it is subjected to criticism and challenged; in response, reasons are offered for believing the claim; these can be challenged in turn, and so on. Together, scientists arrive at a consensus on satisfactory reasons and claims.

4.8 Problems with traditional empiricism

Traditional empiricist philosophies of science do not challenge these facts. Instead, they maintain that, although scientists in fact work this way, the distinction between scientific knowledge and mere opinion can be captured by abstracting a pattern of justification from the messy details of their everyday work. And case histories of good science can be rationally reconstructed to fit the pattern(s). Some traditional philosophies of scientific work picture scientists as making logical connections between data and a hypothesis or theory, and attempt to determine what the logical relationship is. Early examples include Hempel's Hypothetico-Deductive Model of Confirmation, Popper's Falsificationist Model, and, more recently, Laudan's Reticulation Model. On many empiricist accounts, the data alone, or data plus epistemic virtues (constitutive values) and/or other truth-conducive considerations, determine the best hypotheses, models, and theories; or data plus epistemic considerations lead to a good fit between a hypothesis or theory and the evidence for it. On all these accounts, moral, social, or political assumptions (contextual values) have no place in the "context of justification" of hypotheses, models, or theories. If they find their way in, the results are bad science. In the "context of discovery," however, social values are unavoidable, i.e. social interests and values, or even personal whims or dreams, can give rise to hypotheses. But the work of proving hypotheses must be "value-free."

Even recent developments, far from the views of Hempel and Popper, can be understood as belonging to this empiricist tradition inasmuch as they are still attempting to capture the special elements of scientific rationality that make the knowledge claims of science superior to other claimants. For example, Philip Kitcher (1993) holds, roughly speaking, that, since the goal of science is to achieve a unified account of the structure of nature, i.e. the natural kinds and objective dependencies (including causal relations) in nature (Kitcher assumes that nature has a structure, a set of natural kinds and a set of objective dependencies), the consensus practices of scientists working in a domain are rational if and only if they contribute to a unified account of the structure of nature in that domain. In (1993) Kitcher still assumes that social values and interests are distorting biases and the job of philosophers of science is to discover the elements of epistemically well-ordered science. These should include social structures in science insuring that when scientists act so as to maximize their own interests, they are also maximizing significant truth, i.e. accounts of the structure of nature. (See Longino 2002: 51–76 for her criticisms of Kitcher 1993. Although Kitcher (2001) argues that social values must help determine scientific research agendas, i.e. which problems scientists work on and which ones they do not, he still does not depart from the empiricist tradition in regarding social values as distorting biases in the context of justification. The determination of research agendas belongs in the context of discovery, and as long as social values do not affect research while it is being conducted, i.e. the context of justification, science remains "value-free.")

As we have seen, Longino maintains, contrary to traditional empiricism, that the objects, events, and states of affairs providing evidence for hypotheses, models, and theories do not carry labels showing what they are evidence for. Instead, how one determines evidential relevance always depends on one's background beliefs or assumptions. Thus, social values and interests enter the context of justification as background assumptions that, along with many constitutive assumptions, are necessary for deciding when *e* is good evidence for h (Longino 1990: 45 and 43).

4.9 Underdetermination by evidence

This insight is also captured in Longino's argument that hypotheses, models, and theories are logically underdetermined by the evidence/data used to support them. Thus, a set of observations provides evidence for a hypothesis only when scientists make certain assumptions. If the scientists make different assumptions, then the same observations can provide evidence for a different, perhaps conflicting, hypothesis. For example, those seventeenth-century astronomers who assumed that the stars are not very far away took the failure to observe stellar parallax as evidence that the earth does not move around the sun. But those astronomers who assumed that the earth moves around the sun took the same fact, failure to observe stellar parallax, as evidence that the stars are very far away (Anderson 1995b: 28).

Constitutive values can be explicitly called upon or assumed by scientists when the evidence does not determine between two competing hypotheses, and empiricist philosophers including Quine and Kuhn have argued that they should be used in this way. But neither philosophers of science nor scientists have ranked the constitutive values or given definitive interpretations of them. For example, conservatism is usually a good thing for hypotheses, i.e. a new hypothesis should be consistent with/fit with current theories and models. But not always. If conservatism were always highly valued, the history of science would have seen fewer new theories. And if it were valued over (some interpretations of) simplicity, astronomers might have delayed the shift from geocentrism to heliocentrism.

The epistemological underdetermination problem arose when philosophers were forced to give up the logical empiricist view that all meaningful theoretical statements can be fully translated into theory-neutral observation statements. Without this possibility, the nature of the relationship between observation and theory becomes problematic, particularly since observations are "theory-laden." There is no theory-neutral set of statements describing the evidence. Thus, it is not possible to use one set of unequivocal observation statements to decide between two conflicting theories. Instead, observations rest on background assumptions that include theoretical commitments; for example, astronomical observations using a telescope assume that the "telescope is transmitting light from the heavens and not producing images internally, or not systematically distorting the light it receives" (Longino 2002: 126 and Duhem 1954, cited in Longino 2002). There is a logical gap between evidence, e.g. observations that postulate certain entities such as visual images, and hypotheses that postulate different entities and processes to explain the observations, e.g. light emitted from far distant stars. The gap is filled by background assumptions including both substantive and methodological hypotheses (Longino 2002: 125–7).

Scientists, as Longino notes, usually don't articulate these assumptions; they assume them. But when these assumptions are articulated, they, too, must be justified and, when they are justified by evidence, the same underdetermination problem besets them. Although they are theory-laden in this way, data can still serve as evidence for a hypothesis as long as the theory to which the hypothesis belongs and the theory behind the data are not the same. If the data and the hypothesis belong to the same theory, appealing to the data would beg the question in favor of the hypothesis.

Acting together, scientists do, in fact, make such background assumptions and decide to take data as evidence for hypotheses, models, and theories. But, as we have seen, traditional philosophers are worried that, without a logical or other specifiable general pattern or schema, their decisions might be arbitrary. Thus, without such a pattern, scientists' rejection of creationism is just what fundamentalists claim it is: a decision by the socially powerful based ultimately on secular humanist beliefs, not based solely upon solid, determining evidence.

4.10 Social norms

Longino's solution to the problem is to offer a normative theory of scientific knowledge that recognizes the deeply social nature of knowledge. The norms set out in her theory apply to the social practices scientists engage in when they produce knowledge. Thus, she proposes, in addition to traditional empiricist norms such as engaging in *valid reasoning* and providing *empirically adequate* hypotheses and models, that the discursive interactions through which scientists produce knowledge should also meet the following requirements:

1   Venues. "There must be publicly recognized forums for the criticism of evidence, of methods, and of assumptions and reasoning." Criticism should be given nearly the same weight and presented in the same venues as original research. This requirement is difficult to satisfy because of "limitations of space, the relation of scientific research to production and commerce whose consequence is privatization of information and ideas, and the understanding of research as the generation of positive results. [So negative criticisms are not highly valued as research.]"

2   Uptake. "There must be uptake of criticism," not just toleration of it. Although scientists should not capitulate to criticism, they should participate in critical discussion and, over time, beliefs and theories must change in response to the critical discourse.

3   Public Standards. There must be publicly recognized standards for evaluating theories, hypotheses and observational practices. These standards are subordinated to the overall cognitive aims of the community and so provide markers for distinguishing criticism that is relevant to goals of the community. Standards get their grip because individuals acknowledge the relevance of the standards to the evaluation of cognitive practices in their community of inquiry. And the standards themselves can be criticized and changed in reference to other standards, goals, or values held temporarily constant by the community.

4   Tempered Equality. In order to expose hypotheses to the broadest range of criticism, cognitive communities should be characterized by equality of intellectual authority; but this equality must be tempered or qualified to insure the diversity of perspectives necessary for effective critical discourse. Such consensus as is reached must be the result of critical dialogue in which all relevant perspectives are in fact represented (not just have an equal chance at being represented). Thus, the exclusion of women and minority men from scientific education and

professions is a *cognitive* failure. Feminist science scholars have shown "how assumptions about sex and gender structure a number of research programs in biological, behavioral, and other sciences [Keller 1985, 1992; Bleier 1984; Hubbard 1990; Jordanova 1993; Haraway 1989, 1991b]. Historians and sociologists of racist practices and ideologies have documented the role of racial assumptions in the sciences [Lewontin *et al.* 1984; Proctor 1988; Hubbard 1990; Jordanova 1993; Haraway 1989, 1991b]." To guarantee that the requirement of Tempered Equality is met, a community must "take active steps to ensure that alternative points of view are developed enough to be a source of criticism and new perspectives" (Longino 2002: 129–32). This criterion of good scientific practice would insure that feminist perspectives are among those developed and considered. (Longino holds that these norms are provisional; thus, provisionality is itself a meta-norm that all other norms be provisional. See Longino 1994; and see especially Wylie's consideration of the implications of Longino's research norms and meta-norms for feminist epistemology and philosophy of science in Wylie 1994. Chapter 6 of this book analyses two meta-norms commonly assumed in philosophy of science and the contrary meta-norms adopted in feminist philosophies of science.)

4.11 Situated knowers

Knorr-Cetina and other sociologists and anthropologists of science describe the local, contextual, and social nature of actual scientific work, ranging from the interactions between scientists and funders, between scientists and users of their results, to interactions among scientists inside and outside the laboratory. Philosophers do not usually observe this work or read these accounts; they often use the journal articles and other publications in which results are described, abstracting lines of argument and ultimately abstracting patterns or principles of argument from them. But sociologists of science such as Knorr-Cetina point out that the "context of justification" is in fact characterized by scientists using one another's results. If results are not used (i.e. cited, assumed in experiments, modified, etc.), they "wither from lack of attention and fail to be incorporated in the 'body of scientific knowledge.'" Thus, interactions among scientists in fact constitute or produce the body of scientific knowledge. And scientists select not only which results to incorporate, but also which ideas to test, tools to use, procedures to adopt, and so on. Knorr-Cetina and other sociologists have observed that these selections depend upon local contextual features. Thus Knorr-Cetina notes that the drought and energy crisis in California at the time she was observing scientists there strongly affected the selection of procedures: those "using less water or less energy being favored over those requiring more." The local, contextual nature of the selections that produce scientific knowledge is typified by one researcher's answer when asked how to identify successful ideas: "you try to limit your interest to the idea you know is going to be most productive as quickly as possible within the frame of facilities at hand" (Knorr-Cetina 1981: 75, cited in Longino 2002: 27–8).

Such contextual factors are purged as articles are prepared for publication. Longino summarizes:

> Knorr-Cetina traces a paper from its beginnings in the laboratory, through its fifteen drafts in response to internal and external peer review. The agency of the experimenters is progressively eliminated as is any evidence of choices they made regarding, for example, procedures or reagents. The relation of the work to prior work by others is carefully presented. . . . The article is rewritten to incorporate peer reviewers' suggestions, as well as to transfer responsibility for the conclusions drawn from the researchers to their results. The problem, not the researchers, determines the tools; the results, not the researchers, determine (or "suggest") the conclusions.
>
> (Longino 2002: 28)

Scientists understand that articles are constructed in this way, but philosophers often mistake them for accounts of actual work. This mistake in part lies behind recent philosophical arguments that data alone or data plus epistemic virtues determine hypotheses, so no decisions are being made and, so, no social decisions, i.e. decisions resulting from discursive interactions, negotiations, among scientists (Longino 2002: 28–9).

As we have seen, mainstream sociologists of science emphasize the importance of aspects of the social location of scientists such as "dependence on government agencies and industry for funding, their location in an intellectual lineage, in a family, in a particular research group"

(Longino 2002: 107). Feminist science scholars have emphasized the gender, race, and class locations of scientists as bearing on their choices and decisions.

Recognizing that knowers are differently situated in these ways threatens to make knowledge relative to social location. It would seem that a group of scientists who share a social location, e.g. have the same gender, class, and intellectual lineage, and belong to a particular research group, might very well produce different knowledge from another group of differently situated scientists. It is tempting to conclude that the only way to avoid this relativism is to find "context-independent and value-neutral methodological rules" or algorithms that epistemically equivalent knowers use to distinguish knowledge from opinion. Longino argues, to the contrary, that we need an epistemology for actual, empirical subjects either in addition to or instead of an epistemology for idealized subjects (Longino 2002: 95). The four norms set out above, particularly Tempered Equality, arise from the recognition that knowers are situated and *serve to distinguish knowledge-producing communities from communities that are not.*

4.12 Objectivity

Because what counts as evidence for a hypothesis and how strongly it supports the hypothesis are relative to the background beliefs of the local scientific community, the objectivity assumed to characterize scientific inquiry appears to be in jeopardy. Longino notes that objectivity primarily characterizes modes of inquiry that use non-arbitrary and non-subjective criteria for developing, accepting, and rejecting hypotheses and theories. Such modes of inquiry are known as "scientific method" and stand a fair chance of producing unbiased and unprejudiced knowledge. The question then becomes how this can be insured when evidence is relative to the background assumptions scientists hold.

To answer this question, Longino (1990) argues that scientific knowledge involves such social activities as repeating one another's experiments and subjecting one another's work to peer review. And until the results are absorbed into the ongoing work of the scientific community – as evidenced by the citation, use, and modification of the work – the work cannot be said to constitute "scientific knowledge." Objectivity is, thus, a result of the *social* nature of scientific work: science is not a collection of results from isolated individuals; rather, scientific knowledge is produced through the criticism and modification of individual (and group) results. Here Longino distinguishes three ways in which scientists criticize work: (1) Evidential criticism; that is, assessment of degree of support data provide for a hypothesis, the accuracy of experimental results, and the extent and conditions of experimentation. (2) Conceptual criticism addresses the soundness of a hypothesis, its consistency with accepted theory, and the relevance of the evidence provided for it. And, finally, (3) scientists can be critical of the background assumptions supporting scientific work. Longino notes that objectivity includes blocking the influence of subjective preferences in background beliefs and this can be achieved by articulating these background beliefs and having the scientific community criticize them. Objectivity is, then, the result of intersubjective criticism (Longino 1990: 71–3).

The four criteria Longino presents in (2002), Venues, Uptake, Public Standards, and Tempered Equality, can be seen as enhancing the three ways in which scientists criticize their work, especially their need to subject background assumptions to critical scrutiny. If a local community is not very diverse in its membership, many of its background assumptions are invisible to its members and these common assumptions will not be subjected to critical scrutiny. The community needs a variety of perspectives to insure that its data, hypotheses, models, etc. are intersubjectively invariant. Longino dubs the four criteria "*conditions of effective or transformative criticism*," and they characterize the ideal epistemic community. Real epistemic communities can realize them to a higher or lower degree. But they are among the conditions that distinguish subjective opinion – whether that of an individual or a group – from objective knowledge. Since cognitive practices are social, scientific justification must be social, and, as social, these justificatory practices must be subject to evaluation by norms of social practice distinguishing social interactions that produce knowledge from those that do not (Longino 2002: 134).

4.13 Definitions of knowledge

The basic empiricist norm requires that a hypothesis, model, or theory be justified by data in order to be known. In her new (2002) definitions of knowledge, Longino adds to the basic norm her four social norms applying to the discursive interactions that constitute knowledge. Thus, she proposes a new definition of justification or "epistemic acceptability":

EA Some content A is *epistemically acceptable* in community C at time t if A is or is supported by data *d* evident to C at t in light of reasoning and background assumptions which have survived critical scrutiny from as many perspectives as are available to C at t, and C is characterized by venues for criticism, uptake of criticism, public standards, and tempered equality of intellectual authority.

With this definition of epistemic acceptability in mind, knowledge as *content* is defined as

KC A given content, A, accepted by members of C counts as *knowledge* for C if A conforms to its intended object(s) (sufficiently to enable members of C to carry out their projects with respect to that/those objects(s)) and A is epistemically acceptable in C.

What counts as content is not limited to hypotheses or other linguistic entities, but can include anything scientists can be said to know, e.g. models, and can be embodied not only in mental representations, journals, books, discs or the internet, but also in diagrams, three-dimensional models, instruments, or products such as electron micrographs or PET scans.

(Longino 2002: 135–6)

To understand her definitions, we must understand Longino's new success category, *conformation*. Traditionally, philosophers have defined empirical "knowledge" in terms of propositions that correspond to the facts or, more recently, those which cohere in the proper way with other propositions. The propositions that successfully correspond or cohere are said to be "true." Longino points out that these views are fine for propositional knowledge of such facts as that the cat is on the roof, but not for cases in which the corresponding fact is not easily ascertainable. Nancy Cartwright has argued that the laws of physics lie, by which she means that they are not true, i.e. they do not correspond to the facts, unless they are qualified by a number of clauses; for example, Boyle's Law of Gases, $Pr = VT$, which says that the pressure, volume, and temperature of a gas are proportional to one another, is false. It is not true of any gas in the world. It is often called the "Ideal Gas Law" because it is not true unless it is restricted to ideal, not real, gases. But once it is restricted to ideal gases, the law loses its ability to explain real gases. To explain a real gas, it must be subjected to many, many qualifications. Nevertheless, the Ideal Gas Law is successful and very useful, so its success cannot rest upon its *correspondence to facts* about gases (Longino 2002: 109–10).

Longino reminds us that scientific knowledge is not comprised of lots of independent propositions; instead, each belongs to a network of related representations. Boyle's Law belongs to a network of measurements of temperature, volumes, etc. as well as other laws, e.g. about the nature of heat and so on. Even the measurement of the temperature of a room, e.g. "this room is 68° Fahrenheit," is not, strictly speaking, true. With a precise thermometer, we will find variations of temperature in different parts of the room. Yet, we can determine criteria for the adequacy of such claims and, ultimately, for Boyle's Law. These criteria depend on our purposes; if we want to keep our house plants healthy, the reading from a good hardware-store thermometer is adequate. Thus, "it is 68 ° in this room" *conforms* "to the degree required for the purposes at hand" (Longino 2002: 113). If we want the interior of our container to reach absolute zero Kelvin, we must use better probes and decide when we have made enough attempts with them to measure the temperature inside the apparatus. Our knowledge that the interior is 4° Kelvin belongs in a large network of propositions about our equipment, about heat, and so on. No *isolated* proposition, e.g. "the interior is 4° Kelvin" conforms to the facts. That it conforms at all depends, as we see in our examples, upon other propositions, including laws such as Boyle's Law.

If we do not limit scientific knowledge to sets of propositions or sentences, and recognize that it can be captured in such things as diagrams and images, we might join Longino and others in taking models as the general category for the *content* of scientific knowledge. Thus, a successful model is not true, but its structure, or part of it, may be identical to some structure in the world. The relation between such a model and the world would be isomorphic. Maps are good examples of such models; these can be isomorphic or homomorphic. The point here is that much scientific knowledge is expressed or represented in non-propositional content and our conception of a scientific theory must include all scientific knowledge, and, in many cases, it is stretching the notion of "truth" to argue that all scientific knowledge must be "true."

Thus, Longino mentions a number of alternatives to "true" for expressing representational success: isomorphism, homomorphism, truth, approximation, fit, and similarity. These are modes of *conformation*. On this view, mapping is a pretty good model for scientific knowledge because the success of a map depends on the purposes for which it is used as well as on our mapping conventions and, after all, on the terrain being mapped. There are also degrees of conformation, i.e. how well the map conforms, and respects of conformation, e.g. the Peterson projection maps with respect to the relative landmass of the continents. And once the degrees and respects are specified, whether the map is successful depends upon the terrain of the earth (Longino 2002: 115).

> "Conformation" covers a family of success categories; the particular kind of success required depends upon features of the content – whether it is a singular statement, an abstract physical principle, statistical claim, diagram, schema, equation, etc. Among the categories of representational success, truth is a limiting concept – here degree and respect fall away. The rest are subject to degree depending upon the purposes and methods of the scientific work under way.
>
> (Longino 2002: 121 and 117)

Longino's second definition of knowledge concerns *knowledge-productive practices*:

> KP Processes and practices of content construction and acceptance, such as those comprised in the categories of observation and reasoning, are *knowledge-productive practices* in C, if, when engaged in by members of C, they tend to result in the production or adoption of epistemically acceptable content that conforms to its (intended) objects sufficiently to enable members of C to carry out their projects with respect to those objects.
>
> (Longino 2002: 137)

This definition is a concise summary of Longino's normative suggestions for good knowledge production.

And turning to the traditional concern of philosophers, knowledge as an attribute of a knower, Longino offers the following definition:

KS S *knows that p if*
  1   *s accepts p,*

2   p (or p conforms to its intended object sufficiently to enable members of C to carry out their projects with respect to those objects)

3   S's response to contextually appropriate criticism of p or of S's accepting p is or would be epistemically acceptable in C (i.e. S's response does or would satisfy standards adopted by C, and would itself be so evaluated by the relevant sub-group of C, in situations characterized by the conditions of effective criticism) (Longino 2002: 138).

This definition of knowledge is similar to the one with which students of philosophy are most familiar:

1   S believes that p (the knower, S, has a representation, here a belief that p)

2   p is true (the belief accurately corresponds to the object)

3   S is justified in believing that p (S has good reasons, or the proper warrant, for believing that p).

On these definitions, knowledge is a three-term relation among the knower, the representation or content, and the object of knowledge, and is attributed to the knower, S. On the traditional definition, knowledge is defined as a type of belief and belief is a state of an individual person, S, the knower. Longino leaves open what it is for S to accept a content, p.

Radical relativism holds that justification and knowledge are relative to the individual at any single moment, thus the opinions of all individuals at each moment are epistemically equivalent. Longino's definitions show that knowledge is relative to communities, but not every community is knowledge-producing; only those engaged in knowledge-productive practices as defined in KP above are producing knowledge. The rest are producing opinions. Thus, scientific knowledge is produced by scientific communities, not isolated individuals, and it must be epistemically acceptable, i.e. be supported by the data and have survived critical scrutiny in accordance with the norms of effective criticism, and it must conform to its intended objects, e.g. observable and unobservable facts, sufficiently for scientists to carry out their projects on those objects (Longino 2002: 138).

4.14 Who knows?

Who is "S"? On Longino's social account of knowledge, nobody knows a hypothesis, model, etc. until the community, C, has processed it properly. The isolated scientist cannot produce knowledge because producing knowledge requires practices, engaged in by members of C, that tend to produce epistemically acceptable content, i.e. content supported not only by the data, but also by reasoning and background assumptions that have survived the critical scrutiny (characterized by the four criteria) of other scientists (Longino 2002: 122 and 145).

Turning to knowledge as an attribute, certainly a group of scientists can be said to know something, e.g. "Does the phage group at Stanford know this?" "Yes, they do." The phage group may not have produced the knowledge, but if it is epistemically acceptable and it conforms to its intended object sufficiently, and is transmitted to the group and they have satisfactorily responded to challenges to it (or they could do if challenged), they know it. Similarly, any individual who was part of the group producing the knowledge or to whom the knowledge has been transmitted (maybe she attended a seminar on it or read an article), if she can satisfactorily answer challenges to it in an appropriate situation, knows it. (An appropriate situation is one characterized by the four criteria of successful criticism.)

But an epistemically isolated individual scientist does not know anything. As we have seen, she cannot produce knowledge alone. But suppose she thought up the original model. She recognized its importance right away; she ran a few tests on it over the weekend when nobody was around. She said to herself, "I know this is the right model!" Didn't she know then, before her colleagues accepted it? On a social account of knowledge, the individual can, of course, be said to have "known all along," but this is because it turned out that her model was epistemically acceptable to her colleagues in C and conformed to its intended object well enough for them to carry out their projects using it and she could successfully respond to challenges to it. She might have thought up the model alone and run a few tests on it, but it is not part of the body of knowledge until it is deemed epistemically acceptable by her colleagues and has been subjected to criticism by a diverse group of them. We can see, then, that the S who knows can be an individual, but not an epistemically isolated individual. According to Longino, our scientist can believe that her model is right, but she cannot know it until it has been through the proper social processes. (This does not preclude our saying, retrospectively – after it has been through the proper social processes – that "she knew it all along.")

It is because an individual (though not an epistemically isolated individual) has what Longino dubs "doxastic autonomy," i.e. her beliefs need not be constrained by discursive interaction in the way that knowledge must, that individuals can come up with cogent criticisms of the hypotheses and models put forward in her scientific community. The ideal scientific community, C, is comprised of individuals with different points of view: different "attitudes, opinions, evaluative categories, social location and spatiotemporal orientation." Each of these individuals belongs to other cognitive communities within and without the world of science and the insights gained in other communities function as epistemic resources in C. If S can transfer these other insights in terms of the standards of C, she can introduce yet another test the hypothesis must pass. She can even criticize the very standards of C themselves "by exploiting the inevitable tensions within any given set" of standards. And she is more likely to do this if she compares the standards of C with standards in her other communities (Longino 2002: 154–5).[1]

4.15 A dilemma

Longino and many other feminist philosophers of science have endorsed the view that knowledge is local in the ways discussed above. Thus, a good epistemology must be one that describes local knowledge practices and sets out the norms inherent in those practices, norms that distinguish scientific knowledge from opinion. What, then, asks philosopher Robert Audi, is the status of the four norms of transformative criticism Longino sets out? Are they merely a local community's conditions of criticism? If so, they are binding only on members of that scientific community and do not have a general claim on all scientists in every scientific community. "But if they are general," Longino notes, "then their proposal as normative contravenes the view that normative epistemology is local." To solve the dilemma, Longino proposes that the four norms are an explication of "objectivity." Thus, they are produced by philosophers who are clarifying a shared concept of knowledge and a shared intuitive distinction between knowledge and opinion. They will, then, hold for everyone who shares the same concepts of objectivity and of knowledge, and makes this distinction between knowledge and opinion. If there is a distinction between knowledge and opinion, and objectivity is part of what distinguishes it from opinion, then "knowledge" is a normative concept. If Longino's four norms successfully capture part of what it means to be objective and to know something, then they hold for all who use this concept of knowledge and who wish to be objective and to produce knowledge (or recognize it).

> Those who reject the four conditions have a different concept of knowledge or a concept of something else. So there is a sense in which even the general conditions are local, just less local than particular norms adopted in particular communities in conformity with the general conditions.
>
> (Longino 2002: 174)

Although still too relativist for some philosophers, Longino's is not a pernicious relativism and, on her account, it follows from the demand that epistemology of science be naturalized, empirically adequate to the history, including the current practices, of science.

4.16 Conclusion

The feminist view that androcentric and sexist values and interests appear in the "context of justification," i.e. in the "content" of scientific work and the knowledge produced in that work, is supported by Longino's epistemology of science. Her model of the connections among evidence, background assumptions, and hypotheses shows us how androcentric and sexist values appear in results of good, standard scientific work. It also supports the view of some feminists that since social values enter scientific reasoning, and yet the hypotheses, models, and theories produced are *empirically adequate and constitutively virtuous*, feminist scientists can use feminist background assumptions and still produce good science. By the same token, her account also supports the feminist view that objectivity can be maintained while admitting that social and political values operate in standard work and that feminist and non-feminist scientists bring different points of view and assumptions to research. Indeed, objectivity is enhanced on her model by the diversity of scientists and of research programs. Finally, Longino's account supports the feminist desire that, where appropriate, feminist research programs be pursued along with mainstream research programs and the feminist view that the overall result will be better science.

5

STANDPOINT EPISTEMOLOGIES
OF SCIENCE

There are many standpoint epistemologies and standpoint epistemologies of science. In this chapter, we will set out the most influential, that of Sandra Harding, who is one of the founders of feminist philosophy of science. We will also take a brief look at the most recent philosophical development of standpoint epistemology of science, that of Alison Wylie.

Harding has continued to elaborate her standpoint epistemology of science since its inception in the late 1970s. We will not attempt to trace these modifications, many of them merely shifts in emphasis; instead, we will set out the central theses and arguments as they appear in her recent work.

5.1 What is a standpoint?

A standpoint is an achievement, the result of analysis by more than one person who, in the first instance, occupy a particular location in a political order. On this understanding, the social/political organization of societies includes gender relations, race and ethnic relations, class relations, and many other relationships among people. And as a matter of fact, in most contemporary societies, these are hierarchical relations. Different groups of people can be located at the various intersections of two or more of these hierarchies, and we may refer to these intersections as "social locations." Thus, we can "locate" white working-class women, middle-class Chicanas, white upper-class men, etc.

A standpoint is not the same as a viewpoint or a perspective, for any group of people occupying a common social location (e.g. Puerto Rican immigrants to the US mainland) may unreflectively hold a point of view or perspective. Such a perspective may be typical of people occupying that location but the perspective is not a standpoint. A standpoint arises when people occupying a subordinate social location engage in political struggle to change the conditions of their lives and so engage in an analysis of these conditions in order to change them. Thus, a standpoint is an achievement that is "struggled for." We can find a recent example in consciousness-raising, formalized among some groups of women in the 1960s and 1970s, which contributed to the activities of the various women's movements at that time.

Sandra Harding describes a standpoint as "an objective position in social relations as articulated through one or another theory or discourse" (Harding 1998: 150). However, although the people who work out a standpoint use a theory or set of discursive tools (however informal the theory might be), the standpoint of the group cannot be predicted by the theory since the standpoint depends upon the social facts about the group's location. Moreover, dominant groups as well as subordinate or marginalized groups have discursive frameworks, conceptual schemes, and epistemes within which they understand nature and social relations, and, standpoint theorists argue, dominant groups use their frameworks to structure social relations. Usually, though not always, these conceptual frameworks are held unreflectively; they are not often the result of analysis intended to maintain dominance.

Harding offers not a particular standpoint analysis, but a standpoint theory. Though she suggests that a standpoint theory can function in many ways, as

> a philosophy of knowledge, a philosophy of science, a sociology of knowledge, a moral/political advocacy of the expansion of democratic rights to participate in making the social decisions that will affect one's life, and a proposed research method for the natural and social sciences,

we will approach standpoint theory as a contribution to the epistemology and philosophy of the sciences. This is in keeping with Harding's view that standpoint theories have been especially useful in accounting for

> differences in patterns of knowledge and ignorance created by political relations. . . . The dominant groups in such political relations produce conceptual frameworks in public policy and research disciplines that value the local knowledge that their own

activities and interests make reasonable to them, while devaluing and conceptually suppressing the patterns of knowledge and competing conceptual frameworks that emerge from the activities and interests of the groups disadvantaged by the power of the dominant groups.

> (Harding 1998: 160 and 106n4, citing Collins 1991, Harding 1986 and 1991, Hartsock 1983, and Smith 1987 and 1990)

These competing patterns of knowledge include knowledge of the natural world as well as knowledge of the social world.

5.2 Roots in feminism and Marxism

The standpoint theories with which Harding's belongs originated in the 1970s as a number of feminist thinkers independently found Marxist insights useful for understanding institutionalized relations between men and women, and for understanding the very different accounts of these relations produced by feminists, on the one hand, and androcentric thinkers, on the other. (Different beliefs include, for example, the belief that it is natural for women to care for children and mates, i.e. cook, clean, and nurse them when they are sick, and natural for men to engage in public-sphere activities, especially economic ones. From a feminist standpoint none of these activities is natural to either gender and could be done by men or by women; moreover, the belief is sexist inasmuch as it serves to legitimate economic and other social practices that advantage men over women. The belief is often unreflectively taken to entail that women who engage in both activities are "failures," i.e. not *good* mothers and wives. Thus, for example, juvenile delinquency is believed to be the result of the working mothers' failures due in large measure to the fact that they are not at home with their children. The public stereotype of this mother is a black woman who is not feminine.) In a discussion of the origins of feminist standpoint theory, Harding notes that these feminist standpoint thinkers recognized that male supremacy shapes social life and, as it does so, it also shapes the production of knowledge; in turn, the knowledge produced reinforces male supremacy and male power to shape social life.

Early on, standpoint theorists used the notion of "ideology" to explain dominant accounts of relations of gender, race, and class, and to show how groups with different gender, class, and racial locations *tend* to produce different accounts of nature and social relations (Harding 1997: 384). According to classical Marxism, an ideology is a set of beliefs about social life that misrepresents or distorts the world; and since economic and social conditions produce and reproduce these beliefs, and the beliefs, in turn, legitimate and make the economic and social system seem "natural," the distorted beliefs will not change until the proletariat, which has an interest in understanding the world as it really is, achieves class consciousness and acts to change economic and social relations (Bakhurst 1994: 192). We can see how standpoint theorists could generalize from the proletariat, the economically subordinated group under capitalism, to women as the subordinated group under patriarchy, to people of color as the subordinated group under white supremacy, etc. In each case, dominant "ideologies" or accounts will legitimate, make seem natural and normal, economic, gender, and racial hierarchies. As Harding explains,

> in societies stratified by race, ethnicity, class, gender, sexuality, or some other such politics shaping the very structure of a society, the *activities* of those at the top both organize and set limits on what persons who perform such activities can understand about themselves and the world around them.
>
> (Harding 1993: 54)

In each case, the subordinated group, the proletariat, women, and people of color, respectively, has an interest in understanding the world as it really is so that they can change it.

The observation that different "locations" in gender, racial, and class relations "tend to generate distinctive accounts of nature and social relations" arises from standpoint theories' roots in

> Hegel's reflections on what can be known about the master/ slave relationship from the standpoint of the slave's life in contrast to the far more distorted understanding of it available from the perspective of the master's life. From the perspective of the master's activities, everything the slave does appears to be the consequence either of the master's will or of the slave's lazy and brutish nature. The slave does not appear fully human. However, from the standpoint of the slave's activities, one can see her smiling at the master when she in fact wishes to kill, playing

lazy as the only form of resistance she can get away with, and scheming with the slave community to escape. The slave can be seen as fully human.

(Harding 1998: 149)

We see that the unreflective master accounts for the slave's resistance as laziness, and, unless the slave is conscious of her resistance and thinks about the social relationship she is trapped in, she might accept this view, this "ideology," of herself. Once she achieves consciousness of these relations, however, she has a different account of her resistance.

Harding explains that feminist standpoint theorists addressing the sciences in the 1970s and early 1980s used Marxist epistemologies because only "Marxian epistemology/sociology of knowledge provided . . . resources powerful enough to counter" both major alternatives: rationalist/ empiricist (positivist) and "interpretationist" epistemologies and methodologies (Harding 1997: 383). One of Marxist epistemology's most powerful resources for feminist standpoint theorists was the insight that the material conditions of peoples' lives can actually shape their understanding of the social and natural world. Their knowledge is "socially constructed." To what extent and in what ways it is socially constructed remains in dispute – not just among Marxists – today. As the work of post-structuralists, as well as anti-racist and post-colonialist thinkers became available, Harding used it to develop "post-Marxist" standpoint theory.

5.3 The agent of knowledge

One important difference between the standpoint of the proletarian postulated by Marxists and the standpoint of marginalized people postulated by feminist standpoint theorists is that the proletarian standpoint was conceived as disinterested, objective, having no point of view, no cultural identity; the proletarian was the unitary, universal knower equivalent to "rational man" in social contract theory and in many ways to the traditional empiricist agent of knowledge. This agent of knowledge, best exemplified by the scientist, is supposed to be objective in the sense that, *as a scientist*, he has no point of view and is "universal" precisely because, having no point of view, he knows in the same way that anyone having no point of view would know; all agents of knowledge are the same in this respect. Any cultural, gender, racial, or other social differences among agents are epistemically irrelevant; hence, the epistemic agent is homogeneous and unitary. It follows that the knowledge produced is universal: the same for all knowers. In the same way, knowledge is initially discovered by internally homogeneous individuals and internally homogeneous groups of individuals, "not by culturally specific societies or subgroups in a society such as a certain class or gender or race." Finally, the agent of knowledge differs from the object known (even when the object is a person or group) because the agent *as knower* is "disembodied," i.e. his culture, history, gender, etc. are irrelevant, whereas "the objects [including people as objects of investigation] whose properties scientific knowledge describes and explains . . . are determinate in space and time." Their location, history, etc. explain their properties and actions (Harding 1993: 63–5).

Although feminist standpoint theory inflected by Marxist epistemology was set up in opposition to traditional empiricist epistemologies, some early standpoint theories were understood to share some of the modernist assumptions common to classical Marxism, positivist and post-positivist epistemologies. Thus, early feminist standpoint theories were thought to treat women as sharing a standpoint, as an internally homogeneous and unitary group. (As *women*, all women have the same needs and interests.) However, feminist thinkers recognized that women occupy different social locations (the classification "woman" disguises the different locations, needs, and interests of real women) and this recognition led to a very different, postmodern feminist understanding of knowing agents. For feminist standpoint theory, in contrast to the modernist view, knowing agents are local and heterogeneous. This is because agents *as knowers* are "embodied," having specific gender, racial, class, historical, and cultural locations that shape the content of their thought. Thus Harding argues not only that "the 'scientific world view' is, in fact, a view of (dominant groups in) modern Western societies, as the histories of science proudly point out," but also that the content of scientific thought is shaped (but not *determined*) by its "historical location." It follows that agents of knowledge are fundamentally the same as objects of knowledge in the sense that "the same kinds of social forces that shape objects of knowledge also shape (but do not determine) knowers and their scientific projects" (Harding 1993: 63–4).

To say that knowers are "shaped by social forces" can be understood in many ways. A classic example would be the belief that individuals are naturally competitive. In a classical Marxian understanding, this belief seems reasonable in a capitalist economic system because capitalism sets the value of commodities, including labor, by their market value. Thus, workers compete against each other for jobs (with some exceptions, e.g. some wartimes such as the Second World War in the US when there were more jobs than workers – though not all wartimes, as we see from the 2003 US war against Iraq when jobs in the US were declining), capitalists compete for markets, etc. Social scientists studying people discover that they behave by and large competitively. The media portray and exhibit this competition; churches, synagogues, and mosques compete for believers and homilies take it for granted that people are competitive; school children compete for grades and are taught that people are competitive; women and men compete for one another's affections; they watch nature programs on TV portraying charismatic megafauna such as rams, lions, and male Hamadryas baboons fighting for access to females; they reasonably conclude that competition for "mates" is natural. Marxian analysts are not surprised, then, that science discovers Darwin's fundamental theory about the natural world – first that individuals, and, later, that species, compete with others for survival – to be the basis for sound theories accounting for the evolution of species. Nor are they surprised when historians of science find that Darwin's insights were enabled by his reading of *social* theory arguing that economic competition among people leads to the survival of the fittest. Darwin and most of us who know on the basis of common sense, social and natural science that competition is natural, can be said to be knowers who are "shaped by social forces."

But how can trees and rocks be shaped by social forces like agents of knowledge? As Harding remarks, "they do not think or carry on any of the other activities that distinguish human communities from other constituents of the world around us." But scientists cannot study rocks, trees, planetary orbits, and electrons as they are, in themselves, "untouched by human concerns." Instead, they study "nature as an object of knowledge." Such objects are already socially constituted in some of the ways humans and social groups are already socially constituted for the social scientist. First, they are social objects inasmuch as they have general cultural meanings for everyone, including members of scientific communities. Second, they are socially constituted through the meanings they have for scientists, i.e. scientists understand them through concepts and principles earlier generations of scientists used to understand them. Contemporary scientists must take as assumptions some of the earlier understandings of these objects even when they criticize others; they could not do science if they did not borrow some past understandings of the objects and processes they study. Third, scientists' very interactions with these objects constitute them; "to treat a piece of nature with respect, violence, degradation, curiosity, or indifference is to participate in culturally constituting such an object of knowledge." Thus, the same sorts of social forces that are shaping agents of knowledge "are also thereby shaping their objects of knowledge" (Harding 1993: 64 and 65).

5.4 Communities as agents of knowledge

Against the traditional empiricist assumption that the individual is the primary epistemic agent, feminist standpoint epistemology holds that communities, not individuals, produce knowledge in the first instance. One scientist's brilliant thought becomes knowledge when it is legitimated and taken up by the scientific community. And there are two senses in which the community as the primary agent of knowledge is heterogeneous as opposed to homogeneous. First, different communities *as epistemic agents* may differ in many ways from one another: a community of middle-class Chicanas differs culturally, by gender, ethnicity, and class, from a community of poor, Anglo women. And while different epistemic communities may produce similar accounts of the same aspects of the natural or social world, they may well produce different or even contradictory accounts. For example, if feminist knowledge begins from women's lives, *ipso facto* it begins from many different lives. And "different women's lives are in important respects opposed to each other," for they include European and African, economically privileged and poor, lesbian and heterosexual, Protestant, Jewish, and Islamic women (Harding 1993: 65). If hypotheses about nature and about the social world are taken from the lives or understandings of all these women, then, as we shall see below, the accounts produced might converge, but they might also contradict one another, even as they are less partial and less distorted than accounts produced from dominant social locations.

Second, communities *as epistemic agents* are internally heterogeneous, multiple, and very likely contradictory or incoherent, not homogeneous, unitary, and coherent. This is because they are constituted by people who differ in many important ways. That is, the people within a community might not share many social locations although they share enough to constitute an epistemic community, e.g. the scientific community of neurobiologists. As we shall see below, this internal heterogeneity can be a resource for science, contributing to better science by allowing for strong objectivity.

Unlike classical Marxists and other arch-realists, Harding is not sanguine about the possibility of seeing the world "as it really is" (assuming that the world itself is in fact coherent). The proletarian as conceived by Marxism is supposed to achieve a universal standpoint by having escaped the false consciousness and ideological mystification brought about through living unreflectively in a world in which "the *activities* of those at the top both organize and set limits on what persons who perform such activities can understand about themselves and the world around them" (Harding 1993: 54). Unlike the proletarian, or the agent of modernist epistemologies, the agent of knowledge as feminist standpoint theory understands them are not disembodied, disinterested, and universal, having no point of view. Thus, although a feminist standpoint is less partial and less false than a masculinist one, it does not escape the influence of social forces altogether. Women's lives, like all lives, are socially constructed discursive formations. As Harding puts the point,

> one's social situation enables and sets limits on what one can know; some social situations – critically unexamined dominant ones – are more limiting than others in this respect, and what makes these situations more limiting is their inability to generate the most critical questions about received belief.
>
> (Harding 1997: 54–5)

Generating critical questions about received belief marks the process of producing and maintaining a standpoint. This critical questioning (including questions about the beliefs constituting a standpoint) requires reflection and constant self-reflection. This can be seen as a requirement for strong objectivity.

5.5 Strong objectivity

When scientific results – accounts, models, theories, etc. – are formally presented by the research groups that have produced them, internal differences and disagreements among the researchers almost always disappear in the formal presentation. The research groups are understood by traditional philosophies of science to have reached consensus about the results of their research. They have done so in part by agreeing to particular methods and agreeing that these methods have been properly practiced, used, carried out, and engaged in. Among the methods are methods for eliminating bias in the collection and interpretation of data. Having methods to eliminate bias makes the research group objective and in this sense their results are objective. The methods insure that any individual scientist's interests, prejudices, and personal values do not bias the results.

However, feminist and other science scholars have noted that traditional standards of objectivity are too weak to identify beliefs, interests, and values widely shared by members of an epistemic community such as a community of scientists in a field of natural or social science. As Harding points out,

> widely held beliefs function as evidence at every stage in scientific inquiry: in the selection of problems, formation of hypotheses, design of research ... collection of data, interpretation and sorting of data, decisions about when to stop research, the way results of research are reported, etc.
>
> (Harding 1993: 69)

Thus, the sciences need stronger standards of objectivity; in particular the practices of the sciences need to be strongly reflexive, finding ways to reveal widespread sexist, racist, class-biased, and Eurocentric cultural beliefs, interests, and values. And given that most scientists are socially dominant men, it is unlikely that they will recognize these unconscious beliefs, interests, and values, but good science requires ways to reveal them (Harding 1993: 69).

Harding has characterized traditional methods as providing "weak objectivity," indicating that they are too weak to uncover widely held interests and values. On the one hand, these methods are too narrow:

> value-free, impartial, dispassionate research has been operationalized to identify and eliminate only those social values and interests that differ among the researchers and critics who are regarded by the scientific community as competent to make such judgments. If this community of "qualified" researchers and critics systematically excludes, for example, all African-Americans and women of

all races and if the larger culture is stratified by race and gender and lacks powerful critiques of this stratification, it is not plausible to imagine that racist and sexist interests and values would be identified within a community of scientists composed entirely of people who benefit – intentionally or not – from institutionalized racism and sexism.

(Harding 1993: 70)

On the other hand, weak objectivity is too broad for it seeks to eliminate all social values and interests including those that can contribute to the production of less partial and distorted beliefs and that can advance democracy. (See Chapter 3 for an example of how feminist values can generate better science.)

In societies stratified by the politics of gender and other categories, many groups of women within marginalized races, ethnicities, classes, and sexualities as well as women within dominant groups can struggle against their oppression and achieve their own standpoint, articulating their own understanding of the natural and social worlds. When they do so, their standpoints can contribute to the strong objectivity of scientific accounts. If the standpoints are used to critique dominant accounts of nature and of the social world, they can reveal hidden androcentric, Eurocentric, or class-based assumptions. It is for this reason that Harding argues on behalf of a more inclusive, more democratic, science: the inclusion of women and men who have standpoints other than the dominant one(s) can help insure the strong objectivity of the sciences.

Only the standpoints of those marginal to these scientific communities are strong enough to identify the culture-wide beliefs, interests, and values of scientific communities. Thus, good science needs more democratic knowledge procedures (Harding 1993: 69). We do not yet have detailed suggestions for such procedures, particularly procedures for involving non-experts in actual research (as opposed to their involvement in science policy decisions).

5.6 Women's standpoints on nature

Central to feminist standpoint epistemology of science is the thesis that "Insofar as gender relations structure social relations, human interactions with nature, and their meanings, there will be distinctive women's standpoints on human environments that both enable and limit human knowledge about and interactions with nature" (Harding 1997: 104).

This claim has three parts. First, it says that social relations, human interactions with nature, and the meanings of both give rise to standpoints on nature; second, that there are distinctive women's standpoints on nature; and, third, that women's standpoints both enable and limit our knowledge of nature. We have discussed the basis for part one of this claim; it is a gender-specific version of the argument that the material conditions of peoples' lives shape their understandings of the world. In this section, we will set out some of the arguments for the second part of this claim and, in the next section, we will turn to the third part.

How is it that women can have different understandings of, different standpoints on, the natural world than men have? Surely women and men inhabit the same natural world and have the same accounts of it. Harding puts forward several arguments indicating many ways in which the standpoints of women are distinctive.

First, regardless of how women and men understand them, many of their biological processes and socially assigned activities differ. Women's reproductive systems differ as well as their biological processes from pregnancy to menopause; but, Harding reminds us, there are many other differences between women and men such as percent of body fat, skeletal construction, and susceptibility to the effects of drugs. Too, in most cultures, women have different interactions with nature than men have because women are assigned different activities: "from tending children, the aged, and sick, to other unpaid domestic labor, local community maintenance, clerical work, factory work, service work . . . gathering food, and maintaining subsistence agriculture, herding, and forestry" (Harding 1997: 97). These distinctive processes and activities mean that women are exposed to different aspects and regularities of nature in ways that can make "some theories more or less plausible" to them than they are to "those who interact only with other environments" (Harding 1997: 96–7).

Second, clearly, interests shape the research programs scientists undertake and the questions they ask of nature. This is seen most obviously when funders drive the questions for research. In this way, different interests shape the results the knowledge science produces. For example, as Harding points out:

Bordering the Atlantic Ocean, one group will want to fish it, another to use it as a coastal highway for local trading, a third to use it for trans-

Atlantic emigration or trading slaves for sugar, a fourth to desalinize it for drinking water, a fifth to use it as a refuse dump, a sixth to use it as a military highway, and a seventh to mine the minerals beneath its floor. These differing interests have created culturally distinctive patterns of knowledge (and ignorance) about this part of nature's regularities and their underlying causal tendencies (Harding 1997: 65).

Since women and men are biologically different in many ways and have systematically different interactions with their natural environments, many of their interests in their bodies and environments will be different. (Harding 1997: 97–9). Thus, women have an interest in both basic and applied research that will contribute to their health, i.e. in knowledge relevant to all of their distinctive biological processes. And whether the research begins from the interests of women's health advocates or from the interests of corporations producing, for example, hormone replacement therapies, will shape the questions asked and the results obtained even when the biological processes studied are "the same." As Harding notes, "research that starts out from women's bodies and interactions with nature, too – not just men's – will arrive at more comprehensive and accurate descriptions and explanations of nature's regularities" (Harding 1998: 97; see Section 8 below for an explanation of "more accurate" and "less false" descriptions and explanations).

Third, historians and philosophers of science have found that scientists use many "discursive resources" as they pursue knowledge of nature. These include (but are certainly not limited to) metaphors, models, and narrative structures, which may be roughly characterized as providing analogies between familiar objects and processes, on the one hand, and new objects, phenomena, and processes under scientific scrutiny, on the other. They also include cultural presuppositions. Discursive resources are useful for scientists inasmuch as they can suggest new ways of looking at phenomena, new ways to extend theories and ways to revise them in the face of recalcitrant or surprising observations. They can be drawn from the culture within which a science is practiced as well as from cultures with which the science interacts, from other disciplines, etc. But they can also limit our understanding of nature when they are, for example, androcentric, drawing on cultural ideals of masculinity (Harding 1997: 68–9 and 99–100).

Bonnie Spanier shows us clear examples of gendered cultural presuppositions and metaphors found in the 1965 edition of a prestigious textbook by Nobel Prize winner, James Watson *et al.*:

Active sperm and passive eggs were alive and well in the outmoded views of fertilization in the first edition of *Molecular Cell Biology*. Fertilization involves the fusion of the head (largely the nucleus) of the sperm with the entire unfertilized egg (with its nucleus and a much larger amount of cytoplasm). Just as scientists have inaccurately credited the ejaculation of sperm and the motility created by the sperm's tail with the power that propels sperm to egg, ignoring the critical role of vaginal contractions and sweeping waves of cilia lining the fallopian tubes, the textbook describes the sperm as the active agent in fertilization. The sperm "penetrates" and is "explosive"; in contrast, the egg membrane "fuses with sperm membrane," with a "depolarization of the egg plasma membrane," and a "rapid release of calcium."

Contrary to the image of the sperm doing all the work by penetrating the egg surface with digestive enzymes packaged in the sperm's acrosomal cap, fusion of the egg and sperm membranes involves the activation of the sperm's enzymes by secretions from the female reproductive tract *and*, in some cases, by the protrusion from the egg's surface of microvilli that draw the sperm into the egg cell. . . .

The fusion of egg and sperm does initiate many changes, as the textbook suggests, but the egg is actively involved in ways not even hinted at. For example, microcinematography of fertilization in some species shows a startling and instantaneous change (it could be called "explosive") in the surface of the egg involving a dramatic rearrangement of the egg cell's surface layers. Textbook language such as "the release of the calcium ions in particular is important in activating the egg for further development" tends to cast the egg in a passive role, yet it is the egg that is releasing the calcium ions at its own surface! . . .

What difference do such distortions arising from cultural biases make for science and for society? Sexualizing sperm cells and superimposing stereotypes of the active male and the passive female distort our understanding of the process of fertilization. The harm done by inaccurately associating gender or stereotypical gender characteristics with molecules, cell organelles, or cells is not insignificant when it leads to ignoring critical data, such as the fertilization cone of certain eggs, which in turn leads to emphasizing studies of the role of sperm, but not of eggs, in fertilization.

(Spanier 1995: 59–61)

And again, in the 1987 edition: "Two Distinct Sexes Are Found in E. coli," the authors inform readers of Molecular Biology of the Gene; they then explicate "sexuality" in this bacterium:

As in higher organisms, there exist male and female cells, but these do not fuse completely, allowing their two sets of chromosomes to intermix and form two complete diploid genomes. Instead, the transfer is always unidirectional, with male chromosomal material moving into female cells; the converse movement of female genes into male cells never occurs. . . . [A]ll the cells in mixed cultures rapidly become male (F+) donor cells.

This male/female designation has been customary in biology textbooks since the 1950s, when scientists found that the single celled bacterium E. coli, cultivated as an experimental organism, sometimes transfers a portion of its genetic material from one cell (which in this case is a whole organism) to another. One form of genetic transfer between two cells is the movement of a tiny circle of DNA, called an F (for "fertility") plasmid, with the aid of a bridge called a pilus. The pilus is a thin protuberance that grows out from the surface of the plasmid containing cell (called F+) and attaches to a cell without a plasmid (called F-). In this process, the plasmid replicates, and one of the resulting two copies ends up in the second cell (now F+ because it contains an F plasmid), while the other copy remains in the original F+ cell. . . .

Scientists labeled the donor and recipient cells "male" and "female," respectively, based on the presence (male) or absence (female) of the plasmid. The language used leaves no doubt about the intended meaning of the relationship, referring to "conjugal unions between male and female cells." Even though the bacteria differ only in the presence or absence of the F plasmid, scientists designate them as different "strains" of E. coli, exaggerating the notion of fundamental difference between males and females.

What's wrong with this designation of "male" and "female" bacteria? The scientific definition of "sex" (exchange of genetic material between organisms) is confused with two cultural meanings of "sex." The first is a sexual act between a male and a female, in which the male is the initiator who makes the sexual act happen and who donates genetic material, with the female as the passive recipient, while the second is the gender designation based on the presence of a male signifier ("his sex" has a double meaning which includes that which proves his sex/gender). The cultural meaning of "sex" as physical/genital intimacy is quite different from the scientific meaning, but scientists have not respected that distinction. Nor does this conflation coincide with the scientific definition of "male" and "female" sexes, which depends on organisms forming gametes equivalent to egg and sperm (which bacteria do not). The designation of male and female strains of E. coli is simply incorrect by scientific definition.

Secondly, espousing stereotypes of the male as active and the female as passive, as well as defining female as absence, are simply sexist. Not only that, but the assumption that sexual interchange occurs only between a male and a female is a heterosexist bias. Ironically, after this "sexual" interaction involving plasmid transfer, both partners are male because they each have an F plasmid! That this "heterosexual" interaction changes the "sex" of one participant is ignored. Despite this unusual result, the authors are silent about cultural implications of this natural phenomenon for transsexuality or homosexuality.

It is striking that this scientifically inaccurate and hetero/sexist sexualizing and genderizing of bacterial cells has remained largely unquestioned even in more progressive texts. The designation continues in common scientific parlance.

(Spanier 1995: 56 and 58)

These two examples show us that discursive resources can be very useful for scientists – in the first example, designation of the sperm as the active male led scientists to focus on the contributions of sperm to fertilization. But, unfortunately, the same resource can be counter-productive – designation of the egg as female led scientists to see it as passive and to ignore or misrepresent its activities and contributions to fertilization. In the second example, we see that insisting on gender metaphors can lead to inaccurate science. The scientific meaning of "sex" (exchange of genetic material between organisms) is conflated with the cultural meaning of "sex" (a sexual act between a male and a female) and the scientific meaning of "male" and "female" (forming gametes equivalent to sperm and egg, respectively) is conflated with the cultural meanings (gender designations based upon the presence or absence of sex/gender signifiers). Since bacteria such as E. *coli* do not form gametes, referring to "male" and "female" "strains" of E. *coli* is scientifically wrong. (It also reinforces the belief that "science proves" that men are active and women are passive.)

When discursive resources have sexual and gendered meanings, women and men have different relationships to them. "Women," Harding argues, "will tend to be more sensitive to cultural metaphors, models, and narratives that devalue womanly beliefs, traits, and behaviors; they more easily detect the distorting components of such models of nature." They may not find beliefs, metaphors, etc. that reflect gender hierarchy as plausible or useful as men tend to and may prefer other cultural resources in their stead. For example, designating bacteria cells containing the F plasmid as "F+" cells instead of "male" cells and those without the F plasmid as "F-" cells instead of "female" cells distinguishes them using different cultural resources, viz. the alphabet and symbols for the mathematical operations, addition and subtraction. Men as well as women can take the standpoint of women and recognize such androcentric discursive elements.

The fourth way in which women's standpoints on human environments can differ from those of men and enable and limit knowledge of and interactions with nature arises from the "gendered organization of scientific work." Harding suggests that women and men tend to organize scientific work differently, including the relations among people in their laboratories, "their choices of scientific projects, and their publishing strategies" (Harding 1998: 100–1).

These organizational differences can make a difference to the results produced. Alison Wylie points out that feminist scrutiny of archaeology as a community and a discipline reveals familiar patterns of differential support, training, and advancement of women as well as gender segregation in the research areas women typically work in. Wylie notes that this research is not usually connected to androcentrism and sexist bias in the content of archaeological accounts. But she points to J.M. Gero's (1993) analysis of Paleoindian research (i.e. on the earliest people who lived in the Americas) as making a strong case that gender segregation and bias in research practices has led to incomplete accounts of Paleoindian culture.

There is a strong pattern of gender segregation in this predominantly male field. Male researchers have focused almost exclusively on stereotypically male activities, concentrating on mammoth and bison kill sites, the remains of technologically sophisticated hunting tools, etc. Women in the field are not found in these core research areas; instead, they work on "domestic" sites, temporary blades, and flake tools associated with women, often focusing on the way their edges wear down. This work has been almost completely ignored, as can be seen through analyzing citation patterns, even though it provides evidence that Paleoindians ate a lot of plants (probably foraged) to complement the Pleistocene mammals they ate. (Gero finds that, in this field, unless women co-author with men, they are cited less frequently than men. Cited in Wylie 2002: 188.)

Based on Gero's argument we can see that androcentric research practices led to ignoring work produced by women researchers and this, in turn, allowed androcentric assumptions to frame "man the (bison/mammoth) hunter" models that have created the central problem for Paleoindian research. The central problem is that "the technology, subsistence activities, social organization, mobility" and the way people inhabited the landscape are all explained primarily in terms of male-associated hunting activities, while the activities of women were not essential to any of these factors. In particular, these models assume that the basic diet of Paleoindians was the meat provided by male hunters, with enough plants to keep the people minimally healthy. But this account generates puzzles when it comes to explaining "what happened to the mammoth hunters when the mammoths went extinct!" Holding to these models, researchers have hypothesized that Paleoindians died out and were followed by people who hunted small game and foraged for plants, or that they "effected a miraculous transformation of their entire form of life as the subsistence base changed." Gero argues that these hypotheses are necessary only because researchers have ignored evidence, produced by the women working on microblades and edge-wear analysis, that Paleoindians depended on much more diverse subsistence strategies than man the hunter models acknowledge (Wylie 2002: 189).

5.7 Women's standpoints as resources for science

There are at least three ways that the standpoints of women can serve as resources for the sciences. First, the experiences and lives of women (and of marginalized men), *as they understand them*, provide scientific problems to be explained and research agendas that differ from the problems that appear in dominant frameworks. For example, teen pregnancy is seen in the currently dominant US framework as a problem; for the conservative members of the dominant framework, it is the result of illicit sex; therefore, the solution to the problem is the abstention of teens from sex. This solution requires social science research to find the best ways to achieve teen abstinence. For the liberal members of the dominant framework, it is the result of ignorance and the inaccessibility of cheap, safe birth control methods; therefore, the solution to the problem is education and accessibility. This solution requires biological research to find safe, economical birth control methods and social science research to find the best ways to educate teens. However, beginning research from the experiences and lives of female teenagers means taking their self-understanding seriously, including their understanding of what their problems are. Let us take up the self-understanding of those young women who see pregnancy as a problem, but do not view sex outside of marriage as illicit and who are not ignorant of birth control methods. For them, the problem is that only condoms are easy to get, but their boyfriends do not enjoy sex using condoms and they want their boyfriends to enjoy sex as much as they do. This framework presents different research questions and requires different solutions from either of the problems and research questions defined by the dominant framework.

Second, the standpoints of women can contribute to strong objectivity. Patricia Hill Collins[1] offers two arguments taken up by Harding to show that and how women's distinctive standpoints can maximize/contribute to the strong objectivity of research, particularly in the social sciences:

(A) Women are "outsiders" or "strangers" as these concepts are understood in sociology and anthropology. The "stranger" who lives among "natives" (here men) – but does not "go native" – is both near and remote as well as concerned and indifferent in ways that allow her to "see patterns of belief or behavior that are hard for those immersed in the culture to detect." Women are strangers in this sense because they are treated as such by "dominant social institutions and conceptual schemes." (But men can learn to see the social order from this standpoint.)

(B) Particular groups of marginalized people are also "outsiders within." Examples include domestic workers, any women working in environments dominated by men, or particular groups of women workers, e.g. social science researchers such as black feminist scholars. Black feminist scholars do "women's work" or "black women's work" and so are not engaged in dominant, i.e. white men's, activities. But they also do "ruling work," inasmuch as "ruling work" includes the production and transmission of knowledge in universities. This combination makes them "outsiders within," and can allow them to see the dissonances and consonances between dominant activities and their own "outsider" community's beliefs. When marginalized outsiders within offer their accounts both of themselves and of those in the "center" or dominant culture, and when these accounts are brought together with accounts of the marginalized and of themselves offered by those in the dominant culture, the resulting conflicts and convergences can help to maximize objectivity and to produce less partial and less distorted accounts (Harding 1991: 124–5 and 131–2, and Collins 1986, §15, cited in Harding 1991). The conflicts are especially important because they allow everyone to see "taken-for-granted" assumptions and to ask whether the assumptions are empirically robust (in Harding's terms, which ones are "less false"; this concept is discussed in the next section).

And as the third way in which the standpoints of women can serve as resources for the sciences, Harding suggests that research begun from the standpoint of women can contribute to the production of accounts of nature and social life that are less distorted and less false. As we have seen, taking seriously women's experiences and lives can lead to new research questions and so to "less partial" knowledge, i.e. knowledge that is more empirically adequate. But Harding argues here that the accounts of nature and social life produced in this way are less distorted and less false. She does not claim that women's standpoints are *ipso facto* epistemically privileged, i.e. that women's standpoints simply give rise to less false hypotheses, accounts, or explanations. Such a claim catches its proponent on the horns of a dilemma: it leads to a form of the metonymic fallacy or to vicious relativism. Any claim that there is one "women's standpoint" and that it is epistemically privileged *vis-à-vis* the standpoint of men quickly runs up against the fact that women differ by culture, race, ethnicity, class, sexuality, and other epistemically salient categories; therefore, there is no one "women's standpoint." Instead, there are many. The claim for one standpoint of all women must, then, commit the fallacy of taking one standpoint, e.g. the standpoint of white middle-class feminists, to be the standpoint of all. On the other hand, if each different group of women has their own standpoint and if each is *equally* epistemically privileged, then there is no way to decide among them when they conflict; each is true. This view is viciously relativist.

5.8 Judgmental relativism and less false theories

To avoid such judgmental relativism without recourse to the metonymic fallacy, Harding maintains that if research begins with questions/hypotheses arising from the lives and standpoints of subordinate groups, and if research is organized in more democratic ways, the results are likely to be "less false" or "more accurate" (Harding 1997: 383). What does Harding mean when she says, for example, that "research that starts out from women's bodies and interactions with nature, too – not just men's – will arrive at more comprehensive and accurate descriptions and explanations of nature's regularities"? (Harding 1998: 97).

Here Harding cites N. Katherine Hayles, who makes a four-fold distinction among true, false, not-true, and not-false. As Hayles sets it out, this view combines aspects of coherence theories of truth and correspondence theories; thus, following Popper, Hayles argues that however much we test a set of beliefs, we can never verify it, though "true" makes sense as a limit. That is, we can conceive of having made all possible tests and having found a theory to stand up to them all. Such a theory would be *true* or "congruent" with all test results. If these results are conceived as corresponding to "how the world is," then our theory would be "congruent" with the world in the sense that it corresponds to the world. But human scientists will never achieve such a theory; instead, we can test a theory and find that it is inconsistent with one or more test results, in which case it is *false*. At best, the theory stands up to our tests, in which case it is consistent with our results and is *not-false*. Those theories which have not been tested and those which are "imperfectly tested" Hayles relegates to the category of *not-true*. They are not *false* or *not-false* because they have neither failed nor withstood any tests. In this view, Realists are those who think that theories which are *not-false* – i.e. have stood up to many tests – are true, i.e. there is some sense in which they correspond to or represent the way the world is (Hayles 1993).

Now we can parse "less false." One set of beliefs about something is less false than another if it is more consistent with test results, i.e. it survives more tests than the other one does or than other ones do. If we factor in significance, it would survive more of the important tests. (Here, of course, we would have to determine criteria for significance. See, for example, Anderson 1995b, Solomon 2001, and Kitcher 1993.) In any case, Harding agrees that

> science never gets us truth. . . . Scientific procedures are supposed to get us claims that are less false than those − and only those − against which they have been tested. . . . Thus, scientific claims are supposed to be held not as true but, only provisionally, as "least false" until counterevidence or a new conceptual framework no longer provides them with the status of "less false."
>
> (Harding 1997: 387)

In a view of science as progressive, science is understood to change over time and to progress, not only in the sense that more theories are found to be less false, but also in the sense that the standards for testing improve. Kuhn's notion of "paradigm shifts" keeps the idea that standards can change, but he also thought that the goals and methods of science can change, and, if they do, the basic entities with which sciences explain phenomena can change. Thus, even if a particular word is carried over from one framework or paradigm to another, e.g. "force" is carried over from classical to quantum mechanics, its referent and so its meaning changes. Thus, it is difficult or impossible to compare theories and to say that a theory from one framework is "less false" than a theory from another framework. The frameworks or paradigms are said to be "incommensurable." Harding rejects the incommensurability of frameworks, arguing instead that science studies show how scientists devise temporary, local strategies, "effective, 'good-enough' translations − pidgin languages − and technical equivalences to get from one conceptual terrain to another and to enable them to work together effectively" (Harding 1998: 171). (Peter Galison conceives these overlapping boundary areas as "trading zones" in which scientists from very different fields develop pidgin languages for common communication. See Galison 1996.) Thus, within a framework and between frameworks, theories can be compared and one found less false than another.

5.9 Pluralism

Nevertheless, Harding holds that there is not always one least false theory. Instead, she says, "many highly useful but conflicting representations can be consistent with 'how the world is,' although none can be uniquely congruent with it" (Harding 1997: 383). Thus, Harding allows for a plurality of theories all of which can be empirically adequate and useful, some of which may be compatible, but some of which may contain knowledge claims that are "fundamentally incompatible" (Harding 1998: 66). How can this be?

This pluralist aspect of Harding's epistemological standpoint theory fits well with a metaphysical pluralist assumption that the world is complex, meaning that things in the world have many, many properties. Thus, there are many ways to classify things. Even when two theories are describing the same individuals, the set of descriptions in each theory may not overlap at all. Or they might have some overlap, i.e. they describe some of the things in some of the same ways. For example, both describe an individual as a "person," but theory A describes her as a "woman" because theory A is a theory accounting for gender discrimination while theory B describes her as "black" because B is a theory of racial discrimination. Nor does Harding need to hold that one of these descriptions is "less false" in the sense that the individual "really" is either a woman or black. She cites John Dupré who argues for "promiscuous realism." This is the view that

> many individuals are objectively members of many individual kinds. Thus, I, for example, am a human, a primate, a male, a philosophy professor, and many other things. All, or at least many, of these are perfectly real kinds; but none of them is the kind to which I belong. Since I deny that any of these kinds is privileged over the others, I must, of course, deny that I have any essential property that determines what kind I really belong to.
>
> (Dupré 1996: 105)

As Dupré suggests, we would choose between theories A and B on the basis of "the goals that motivate our inquiry," e.g. to understand gender discrimination or to understand racial discrimination.

Actually, feminists today would choose theory C, a theory recognizing the intersection of race and gender, and attempting to understand the different ways individuals do and do not suffer discrimination depending upon the combination of their race and gender. Theory C would describe our individual as a "black woman." Theory C is an excellent example of Harding's argument that taking the standpoint of marginalized persons can produce less partial and less false theories. Neither theory A nor theory B will be able to account for the particular discrimination that black women face as *both blacks and women*. Beginning from their standpoint, however, and attending to their experiences enabled many thinkers to reject "pop-bead" accounts of oppression (according to these accounts, oppressions are additive, and, like pop-beads, each is unique, and several can be strung together to account for an individual's experiences). "Pop-bead" or additive accounts of black women's oppression typically assume that accounts of the oppression of black men apply fully to black women and accounts of the oppression of white women apply fully to black women (see Spelman 1988). (We still find newspapers referring to "minorities" as, for example, "blacks and women," leaving the location of black women uncertain.) Clearly, theory C is less false than theories A and B since it will account for experiences that are particular to black women and will not make generalizations about "blacks" or "women" that are false of black women.

Harding argues for theoretical pluralism through an appeal to the underdetermination of theories by evidence. In a discussion of philosophical positions with which she agrees and which underpin her account of objectivity, she mentions these:

> observations are theory-laden; our beliefs form a network such that none are in principle immune from revision; and theories remain underdetermined by any possible collection of evidence for them. There are always many additional possibly plausible hypotheses about any state of affairs that have not yet been proposed, or have been considered but prematurely dismissed, and thus remain untested at any given moment in the history of science. Some small subset of them undoubtedly would fit the existing data just as well as whichever one is favored at present.
>
> (Harding 1998: 126)

Such untested or not-yet-conceived hypotheses and theories can each be empirically adequate or "consistent with nature's order, but no one can be uniquely congruent with it." The different theories might converge, diverge, or conflict (Harding 1998: 120). That is, two theories might converge in the sense that they classify many things in the same ways; they might diverge in the sense that they classify some things in different but not in conflicting ways; or they might classify things in contradictory ways. For example, theories A and B might both be consistent with nature's order and yet conflicting if *both* describe an individual as a "woman," but theory A describes her as "black" because one of her great grandparents was "black" while theory B describes her as "white" because she has fair skin and lives her life as a "white" person. Under promiscuous realism, neither her ancestry nor her skin color is essential to her; therefore, she is not *really* one or the other. (This is not to deny that legislators make laws declaring ancestry to be privileged over skin color and so legally define and categorize people by "race" in order to deny them legal rights. Nor is it to deny that ethnic groups may declare skin color to be privileged over ancestry for purposes of ethnic solidarity.)

5.10 Feminist empiricist standpoint epistemology

We have seen that the threat of judgmental or epistemological relativism arises because standpoint epistemologists embrace four claims. A central tenet of standpoint theory is that knowledge is socially situated; this tenet has two parts (1) that knowledge is produced by social groups; and (2) the social location of an epistemic group shapes its knowledge. The threat of relativism follows because feminist standpoint theorists accept those claims and reject the claims (3) that all women or important groups of women share one or more essential properties (properties that make them all women, or make groups of women, e.g. black women, Jewish women, working-class women, etc. *who they are*) and (4) that these essential features give women or important groups of women epistemic privilege over everybody else. Without epistemic privilege, there appears to be no way to decide between or among conflicting knowledge claims produced by different groups of women.

We have examined Harding's standpoint epistemology and the arguments she puts forward to support her view that different groups of women (among other groups) have not epistemic privilege, but epistemic advantages arising from their distinctive understandings of the world. These distinctive understandings can contribute to the strong objectivity of knowledge, making it less partial and less false in the ways we explained above.

Alison Wylie has outlined a standpoint epistemology that shares the assumptions mentioned above, viz. that knowledge is socially situated, that women do not have essential features as women, and that they do not have general epistemic privilege. But she offers a different account of the epistemic advantage enjoyed by groups of women (and by other epistemic groups, as well). She also offers a very different argument supporting her account.

With Harding, Wylie makes a sharp distinction between social location and standpoint: a social location is structurally defined and the defining structures include (but are not limited to) social institutions, and systemically structured roles and relations, especially relations of production and reproduction. These structures create the material conditions of people's lives and structure their social interactions with one another. If they are robust enough, they can shape and limit what epistemic agents can know. These structures are currently hierarchical and can be understood as constituting a system of power relations resulting in different material conditions, different relations of production and reproduction, different kinds of wage labor, and different kinds of affective labor for those in different social locations. This means that the experiences and understandings of those in different social locations can differ; they can differ not only in the content of their knowledge, but also in what they take knowledge itself to be (Wylie 2003: 31).

Standpoints, on the other hand, "are achieved, struggled for, by epistemic agents critically aware of the conditions under which knowledge is produced and authorized" (ibid.). When some of the people in a social location work out an account of the conditions of their lives and of their pre-standpoint understanding of the world, they have developed a standpoint. The standpoint will also include an account of the conditions giving rise to their standpoint itself. Wylie takes as an example of people working out a standpoint theory feminist philosophers "creating a politically sophisticated, robustly social form of naturalized epistemology and philosophy of science". They ask

> how power relations inflect knowledge: what systematic limitations are imposed by the social location of different classes or

collectivities or groups of knowers, and what features of location or strategies of criticism and inquiry are conducive to understanding this structured epistemic partiality.

(Wylie 2003: 31–2)

As Wylie remarks, in this version of standpoint theory,

it is necessarily an open question what features of location and/or standpoint are relevant to specific epistemic projects. ... [W]e cannot assume that gender location is uniquely or fundamentally important in structuring our understanding or that a feminist standpoint will be the key to understanding the power dynamics that shape what we know.

(Wylie 2003: 32)

Using standpoint theory as part of their research methodology, researchers might ask whether the gender location of the people in a particular study, e.g. their being women, is a significant variable in the inquiry, e.g. into the effects of divorce on the people involved. As feminist standpoint theorists, the researchers will use both their own ("outsider") understanding of gender as well as taking seriously the understanding of their women subjects (the "insiders"). This is part of what is meant by "starting research from women's experiences." But Wylie is careful to point out that other locations, e.g. class and race, could be significant to our researchers' inquiry and other standpoints than feminist ones might be key. Thus, social locations and standpoints are relevant to particular epistemic projects, but not to all.

Do social locations and standpoints ever confer epistemic advantage? Wylie argues that, on one account of *objectivity*, some social locations and standpoints confer contingent epistemic advantage "with respect to particular epistemic projects" (Wylie 2003: 34). What is the account of objectivity here?

Wylie does not disagree with Harding that, in scientific research, the inclusion of people from different social locations and/or with different standpoints can contribute to "strong objectivity," i.e. to uncovering and examining widely held but unexamined assumptions. But she offers a conception of objectivity that is closer to some standard empiricist accounts of it, yet allows her standpoint theory to embrace the epistemic advantage of social locations and standpoints.

For the purposes of standpoint theory, she uses a conception of objectivity as a property of knowledge claims. With Elisabeth Lloyd, she distinguishes objectivity as a property of epistemic agents, as a property of the objects of knowledge, and as a property of knowledge claims. Conventionally, epistemic agents are objective when "they are neutral, dispassionate with regard to a particular subject of inquiry" and objects of knowledge such as facts are objective or constitute "objective reality" when they "are contrasted with ephemeral, subjective constructs; they constitute the 'really real', as Lloyd puts it (1995)" (Wylie 2003: 32–3).

Knowledge claims, Wylie suggests, are objective when they maximize some combination of epistemic virtues. Epistemologists and philosophers of science typically differ over the characteristics that constitute epistemic virtues, but almost all lists include empirical adequacy. A good knowledge claim should be empirically adequate. Wylie parses empirical adequacy as "fidelity to a rich body of localized evidence (empirical depth), or as a capacity to travel (Haraway 1991a) such that the claims in question can extend to a range of domains or applications (empirical breadth)." On some lists, empirical breadth is referred to as "scope" or "generalizability"; so, for example, one form of the Ideal Gas Law, $PV = nRT$, has wide scope or empirical breadth; it applies to most actual gases. But in extreme conditions of pressure and temperature, it does not have great empirical depth; it is not an accurate description of the behavior of actual gases. The behavior of an actual gas under extreme conditions could only be described with a high degree of accuracy by a particular approximation, not by a general law.

Objective knowledge claims have, in addition to empirical adequacy, some combination of such epistemic virtues as "internal coherence, inferential robustness, and consistency with well established collateral bodies of knowledge, as well as explanatory power and a number of other pragmatic and aesthetic virtues," e.g. simplicity. (Wylie 2003: 33) The particular combination of virtues that is maximized depends on the specific epistemic project. Different projects require different combinations of these virtues in order to achieve objective knowledge.

Wylie's account of epistemic advantage can now be formulated as: *depending upon the project, some social locations and standpoints have epistemic advantage inasmuch as they allow an empirical assessment of how likely it is that the knowledge particular knowers produce will not be objective, will fail to maximize the epistemic virtues important for that project.*

But a *standpoint* confers additional epistemic advantage. Because those holding a standpoint are critically conscious of the effects of power relations, that is, the effects of social-political location "on their own understanding and that of (some) others," they have epistemic advantage in assessing "how reliable particular kinds of knowledge are likely to be given the conditions of their production" (Wylie 2003: 34). A social location allows those who inhabit it to assess the likelihood of objectivity of particular claims in epistemic projects carried out by particular epistemic agents. A standpoint allows those who hold it to further assess the effects of social location upon the objectivity of particular claims in epistemic projects carried out by particular epistemic agents.

We can schematically represent these formulations of epistemic advantage as follows:

S claims that p.

S_{St} or S_L claims that p': p claimed by S has likelihood r of failing to be objective with regard to domain D (where D includes p, and p' is an empirical assessment based upon evidence that is salient to S_{St} in virtue of S_{St}'s standpoint and to S_L in virtue of S_L's location).

Additionally, S_{St} claims that p'': given the conditions of knowledge production by S in D, p has likelihood of objectivity r' (and p'' is an empirical assessment based upon evidence that is salient to S_{St} in virtue of S_{St}'s standpoint).

And, finally,

S_{St}' claims that p''': given the conditions of knowledge production by S_{St} in D_{St}, p'' has likelihood of objectivity r' with regard to domain D_{St} (where D_{St} includes p'), and given the conditions of knowledge production by S_L in D_L, p' has likelihood of objectivity r'' with regard to domain D_L (where D_L includes p' and p''' is an empirical assessment based upon evidence that is salient to S_{St} in virtue of S_{St}'s standpoint).

In Chapter 2, Section 5, we saw three examples (Watson and Kennedy 1991, Hastorf 1991, and Brumfiel 1991) of the epistemic advantage social location can confer on a group as it assesses the knowledge claims made by researchers in a particular domain. There are many other examples, all constituting

the rapidly expanding body of research on the "archaeology of gender" that has taken shape in the last decade. It is largely due to women who have focused attention on a range of neglected questions about women and gender, but [many of whom] disavow any affiliation with feminism.

(Wylie 2003: 38 and 1997)

We also found an example of the further epistemic advantage conferred on those who hold a standpoint, here a feminist standpoint. Gero's (1993) analysis of Paleoindian research reveals that "the technology, subsistence activities, social organization, mobility" and the way people inhabited the landscape are all explained primarily in terms of "man the mammoth/ bison hunter models" in which the activities of women were not important to the technology, subsistence activities, social organization, mobility, or the way people inhabited the landscape. Thus, a range of possible (and, in many ways, less problematic) explanatory models have not been considered. Gero further points out the conditions of the production of knowledge in the domain of Paleoindian archaeology that help to account for the neglect of alternative explanatory models: this field, she notes, is marked by striking gender segregation; male researchers in core areas of the field focus almost exclusively on stereotypically male activities and pay little attention to the work of women in the marginal area of edge-wear analysis, even though this work could be useful in solving the central problem in the domain – what happened to the Paleoindians when mammoths went extinct. Her standpoint analysis raises questions – though she does not explicitly state them – about the explanatory adequacy and empirical depth of standard models in the domain.

In all these cases, the agents of knowledge use the same standard of objectivity for their knowledge claims, viz. objective knowledge claims maximize some combination of epistemic virtues depending upon the specific purpose at hand. Men and women archaeologists, including feminists, in a domain such as Paleoindian research decide which virtues are relevant to research projects and they decide on the proper balance among them; each local group negotiates these two aspects of objectivity in the same ways that scientists usually do.

Wylie's version of the epistemic advantage conferred by social location and by standpoint escapes the threat of judgmental relativism inasmuch as all parties to a local project agree to the same standard of objectivity – to the same degree that scientists usually do. (See Solomon 2001 for an excellent discussion of consensus in science.) Wylie does not argue that those inhabiting a social location and those holding a particular standpoint have general epistemic privilege or epistemic advantage in all cases. Her view is that such groups have contingent epistemic advantage, i.e. in some areas, as we have seen, but not in all areas (see also Narayan 1988: 35–6, cited in Wylie 2003). In this way, as she puts the point, "Standpoint theory has the resources to explain how it is that, far from automatically compromising the knowledge produced by a research enterprise, objectivity may be substantially improved by certain kinds of non-neutrality on the part of practitioners" (Wylie 2003: 38).

6

CONCLUDING ARGUMENTS

Can the philosophy of science be value-free?

A feminist philosophy of science is, among other things, one that allows us to see whether and how gender politics influence science. Scientific research is understood to be a paradigm of rational thought and practice, so the question for feminist philosophers of science has been "Can gender values influence scientific research and the research still be rational?" The mark of a feminist philosophy of science is that it offers a positive answer to this question. Most respond, "yes, frequently"; some simply respond, "yes."

Philosophy of science in the twentieth century tried to pinpoint what it is that makes science rational; early efforts focused on scientific method. Logical empiricists and their descendants attempted to reduce scientific method to logic, first to deductive logic and then to inductive logic. Those philosophers assumed a sharp distinction between facts and values, and this assumption produced other distinctions such as the one between sentences expressing cognitively meaningful propositions – expressing facts that could be empirically verified or falsified – and meaningless sentences, including those that express values – supposed immune to empirical verification or falsification. But even philosophers who do not believe the distinction between facts and values is sharp often have a lingering suspicion that, if contextual values influence scientific work, the result is bad science. And many cases of bad science were, indeed, the result of the influence of value assumptions on scientists. Lysenkoism and Nazi science come to mind, and feminists think of Victorian period accounts of the bodies and minds of both European and African women.

As we have seen in the preceding chapters, philosophers of science distinguish three points at which values can influence science: they can influence which hypothesis or theory is chosen for research, how the hypothesis or theory is used after it is established, and, of greatest concern, the actual work establishing the hypothesis or theory. Most science observers are increasingly worried about the influence of values and interests of the institutions financing research upon the choice of hypotheses to be investigated, and many philosophers have done excellent work on this area, the "context of discovery" (see, for example, Kitcher 2001). Some of these values are benign; some are not; but there is wide agreement in philosophy that scientific research is influenced by contextual values in this way. And similar interests and values influence the uses to which scientific results are put. Philosophers have, however, been very concerned with whether contextual values and interests influence the "context of justification."

This concern was pressed by Richard Rudner in his 1953 argument, in the pages of the prestigious *Journal of Philosophy*, that (roughly) once he has determined the probability of a hypothesis, the scientist must still decide whether to accept it. And his decision will take account of how important (i.e. how valuable) it is not to make a mistake. Rudner's example was scientists at a pharmaceutical firm trying to decide whether the probability based on the evidence of a drug's being both safe and effective is high enough to accept the hypothesis that "the drug is safe and effective." In 1956 Richard Jeffrey tried to defend the value neutrality of the scientist by arguing against Rudner that the scientist *qua* scientist does not accept or reject hypotheses, he merely calculates their probabilities! This led to a spate of articles and books trying to distinguish among the scientist's "accepting," "believing," "advocating," "merely entertaining," and "using" hypotheses.

Carl Hempel rejected Jeffrey's attempt as "startling" and pointed out that Jeffrey had failed to answer Rudner's argument that, in calculating the probability of a hypothesis, the scientist must accept hypotheses describing the evidence he uses in his calculations and that this acceptance in turn depends upon how important it is not to make a mistake. In 1981 Hempel (whose Hypothetico-Deductive Model of Confirmation is a standard to which all philosophy of science students are introduced) announced a post-positivist project for philosophy of science as (1) finding a satisfactory theory of probability for determining the probability of hypotheses on the basis of given evidence and (2) formulating rules governing the acceptance of hypotheses once scientists determine their probability on the basis of given evidence (Hempel 1981: 392). He was convinced that there was "something right" about Rudner's argument, and his conviction led him to give up the search for a rule of induction governing scientists when they adopt hypotheses on the basis of logically incomplete evidence. Several problems (including the lottery paradox) induced the recognition (a) that any given hypothesis, h, has a probability, r, only in relation to a particular body of evidence, *e*. Thus, there can be no *general rule or principle* for the belief or acceptance of empirical hypotheses based upon induction. And (b) scientists cannot automatically decide to accept a hypothesis when the evidence for it is greater than .5 or when it has some fixed value greater than .5. Therefore, a proper logic of science must give us "adequate criteria for the rational acceptability of a hypothesis," taking account not only of its probability based on the relevant evidence, "but also of the values attached to avoiding the mistakes of accepting the hypothesis when it is, in fact, false; or of rejecting it when it is true" (Hempel 1981: 393).

Thomas Kuhn (1970) and W.V. Quine (Quine and Ullian 1970) had already presented lists of epistemic values that scientists use or should use to decide among empirically underdetermined hypotheses and theories. (We have seen that empirical underdetermination means different things to different philosophers.) Hempel wanted desiderata that contribute to the objectives or goals of scientific inquiry, and his list includes the theory's being probably true, being rich in informational content, closely fitting experimental data, having predictive power, simplicity, and compatibility with established theories in related fields. He says, "it becomes a truism that replacing a theory by a competing one that better satisfies the desiderata will constitute an improvement of scientific knowledge and will thus be a rational procedure" (Hempel 1981: 404).

Having agreed that values influence the acceptance of scientific theories, Hempel is at pains to show that values influence *only* the acceptance of theories, not their "content." He says,

> Since it is often said that science *presupposes* value judgment, let me stress that epistemic judgments of value do not enter into the *content* of scientific hypotheses or theories; Kepler's laws, for example, do not presuppose or imply any value judgments at all – either epistemic or of other kinds. But epistemic valuation does enter into the *acceptance* of hypotheses or theories in this sense: the assertion that a given hypothesis H is *acceptable* in a given knowl-

edge situation implies that the acceptance of H possesses a greater expectable epistemic value for science than does the acceptance of any rival hypothesis that may be under consideration.

(Hempel 1981: 398)

In stating that values are not part of the "content," he is arguing that values are not part of the meaning of, for example, Kepler's laws; their definition does not include values. But "having greater expectable epistemic value for science than its rivals" is part of the meaning of the assertion that "hypothesis H is acceptable." It is part of the meaning because, Hempel says, asserting "that a given hypothesis H is acceptable" *implies* that the acceptance of H has greater epistemic value for science than accepting any other hypotheses currently being considered. And possessing greater epistemic value for science consists in being more probably true, richer in informational content, more closely fitting experimental data, having greater predictive power, being simpler and more compatible with established theories in related fields than current rivals. It follows that since contextual values are not included in the list of desiderata for choosing among rival theories, i.e. are not epistemic values, then they have no place in theory choice.

The epistemic values are what make the decision to accept a hypothesis or theory a rational decision. Hempel asks, "how could a procedure like the adoption of a hypothesis be qualified as appropriate or rational except in consideration of the objectives of scientific inquiry [which the epistemic values contribute to or comprise]?" (Hempel 1981: 398).

Hempel does not *state* that contextual values ought not have a place in rational theory choice; he simply argues that part of the meaning of a theory's being acceptable is that it maximizes more of the epistemic values which constitute the aim of science, and he does not include contextual values in his list of factors contributing to the aim of science.

Thus, we can see the importance (as well as the historical place) of the arguments of Anderson (Chapter 3), Longino (Chapter 4), and other philosophers that truth, empirical adequacy, empirical success, and the like do not constitute the sole aim of science or, in some cases, any part of the aim of a research project. Scientific research to find a useful general law, such as the Ideal Gas Law, is *certainly rational* even though, as Cartwright argues, the law is not true of any real gases; to explain a real gas, it must be subjected to many, many qualifications – only a very specific description of the behavior of a particular gas in particular conditions is true. But such a description is not generally useful the way the law is. We discussed this point in Chapter 4.

We can also see the importance of Anderson's argument that, not only is truth *not* the sole aim of science, it *could not be* the sole aim. This is because truth alone offers no guide for research. The point of research is not to produce any old collection of truths, but to produce a significant representation of some phenomenon. And significance is determined in almost all cases of scientific research by contextual values and interests, for example by our interest in human health; in finding, for example, a cure for breast cancer. These values and interests do not make the acceptance of a theory about breast cancer irrational; in fact, they are necessary in order for the theory to be a rational account of something (here, breast cancer) instead of a meaningless collection of true sentences.

The efforts of epistemologists and philosophers of science to find a suitably broad yet precise conception of scientific rationality have spanned the twentieth century and continue today. Helen Longino offers us such a conception, as we have seen. She suggests that the status of knowledge requires that it have a claim on what we ought to believe. The four norms she sets out – publicly recognized venues for the criticism of evidence, methods, assumptions, and reasoning; uptake of criticism; publicly recognized standards for evaluating theories, hypotheses, and observational practices; and equality of intellectual authority tempered to insure a diversity of perspectives – are part of any account of how a knowledge claim comes to warrant the status of knowledge (attributed to its content). Scientific success in producing knowledge is not just empirical success, i.e. the production of models and theories that are supported by experiment or observation, and/or are predictive, and/or exhibit some kinds of technological success or some kinds of explanatory success (Solomon 2001). Longino argues that we believe, put our confidence in, the claims of the sciences because they also meet the four norms adequately.

But, of course, the knowledge claims of the sciences, and knowledge claims in general, do not meet the four norms adequately, as Longino knows. This is a problem for Longino, but her particular version of the problem is merely an example of the general problem; every effort to find desiderata of scientific rationality has faced the problem: to what degree should the account of scientific rationality be descriptive and to what degree prescriptive or normative? How close should the fit between a philosophy of science and its data, i.e. the history of science including current science, be? How much rational reconstruction is permissible for a naturalizing philosophy of science?

All philosophies of science engage in some rational reconstruction and so lie on a spectrum between the very normative and the very descriptive or naturalizing. Thus, a very descriptive philosophy of science makes it clear that the boundary between naturalizing philosophy of science and empirical sciences such as the sociology of science is quite permeable, as Nelson points out in her arguments for holism. Miriam Solomon's excellent 2001 book is a prime example. At the normative end of the spectrum we find Hans Reichenbach, who made it clear in the first eight pages of *Experience and Prediction* (1938) that what actually happens in the production of scientific knowledge is *almost* irrelevant to the epistemology of science. Arguably, at the descriptive end of the spectrum we find Miriam Solomon's social empiricism as put forward in her book *Social Empiricism* (2001).

Reichenbach set out an apparatus to deal with the fact that no *actual* cases fit his logic of science; his apparatus primarily serves to reconcile his very normative account of scientific rationality and his assumption that rational science is value-free with the fact that his account fits no actual cases. He coined the phrases, (1) *context of justification* and *context of discovery* to mark the distinction between the way a mathematician as a thinker finds his theorem (context of discovery) and his way of presenting it before a public or the way the physicist publishes "his logical reasoning in the foundation of a new theory" (context of justification). The way they publish, Reichenbach says, "would almost correspond to our concept of (2) rational reconstruction" – which he took up from Rudolf Carnap to delineate the philosopher's logical substitute for the actual processes of thought. He says,

> What epistemology intends is to construct thinking processes in a way in which they *ought* to occur if they are to be ranged in a consistent system; or to construct justifiable sets of operations which can be intercalated between the starting-point and the issue of thought-processes, *replacing the real* intermediate links. Epistemology thus considers a logical substitute rather than real process. For this logical substitute the term *rational reconstruction* has been introduced.
>
> (Reichenbach 1938: 5; first two italics mine)

His apparatus also includes a second pair of distinctions; the (3) *internal relations* of knowledge belong to the "content of knowledge" and the *external relations* do not; instead, they are the ways in which knowledge statements are combined with non-knowledge utterances – which could be anything, including especially expressions of value. Finally, there is a division of intellectual labor between (4) philosophers, who are interested in internal relations, and psychologists or sociologists, who treat external relations, e.g. facts about scientists. The internal structure of knowledge is the system of logical interconnections of thought; and this logical system is a logic of science – not a description of actual scientific thinking. He states,

> the scientific genius has never felt bound to the narrow steps and prescribed courses of logical reasoning. It would be, therefore, a vain attempt to construct a theory of knowledge which is at the same time logically complete and in strict correspondence with the psychological processes of thought.
>
> (Reichenbach 1938: 4–5 and 8)

Returning to our earlier example, faced with Rudner's case in which scientists in a pharmaceutical firm accept the hypothesis that "the drug is safe and effective" after calculating its probability, Jeffrey (1) narrows the context of justification, that is, the work the scientists do to establish a hypothesis, to include only determining the evidence through experimentation and calculating the probability of the truth of the hypothesis. As a philosopher (4), Jeffrey (2) rationally reconstructs the case or gives (3) an internal account of it in such a way that values (constitutive and contextual) do not influence the work of the scientists.

The distinction between the context of justification and the context of discovery is still useful as a way to deal with the influence of contextual values upon the justificatory work of science even when the distinction is not explicitly appealed to. For example, as we pointed out in Chapter 4, Kitcher (2001) argues that social values must help determine scientific research agendas, i.e. which problems scientists work on and which ones they do not; but the determination of research agendas belongs in the context of discovery. He has yet to offer us an account of scientific work in the context of justification that does not treat contextual values as distorting biases.

Many philosophies of science allow us to see the place of contextual values in good scientific research carried out to produce data, interpret it, use it to test a hypothesis or model, etc. We have treated five such feminist philosophies of science in this book. These are motivated in part by the feminist interest in modeling the way in which gender values can influence good science so that we can work to insure that the sciences themselves do not contribute to women's subordination. Other philosophies of science (we might call them "feminist-friendly") also allow us to see whether and how contextual values influence scientific decision-making; these include the social epistemology of Miriam Solomon (2001), the decision theoretic model of Ronald Giere (1988), the model theoretic theory of theories of Bas Van Fraassen (1980), and I have argued that Mary Hesse's (1974) Network Model of Scientific Laws allows us to see cases in this way (Potter 1988, 1995, and 2001).[1] Giere's model of rational scientific theory choice provides a good example of the concern to find a normative account that more closely fits the actual facts about cases which the pre-theoretic intuitions of both scientists and philosophers of science deem good science but which reveal the influence of contextual values. Giere notes that not only do scientists value correct outcomes more than mistakes, but they are also not indifferent to the choice between two correct outcomes. Thus, in choosing one, they use scientific (epistemic) values *and* non-epistemic values. He points to work in the sociology of science showing that scientists use "professional and other broader interests" as well as traditional epistemic values to make their decisions. He says, "Since scientists obviously have both professional *and social interests*, any model of scientific decision making that restricted consideration to some supposed set of 'scientific values' would stand little chance of fitting the actions of real scientists." And again, "As I understand it, a cognitive theory of science need not deny the importance of these other interests. If it did, it could not be an adequate theory of science" (Giere 1988: 163 and 165). That is to say, the theory would fail to be an adequate explanation of the facts about how scientists actually choose among theories.

Note that Giere is moved by careful studies in the sociology of science to produce a normative account of rational scientific theory choice that does not treat scientists as irrational, or failing to meet proper standards of rationality, by rationally reconstructing their work to fit a theory of scientific rationality. Reichenbach actually had to write the sentence quoted above:

the scientific genius has never felt bound to the narrow steps and prescribed courses of logical reasoning. It would be, therefore, a vain attempt to construct a theory of knowledge which is at the same time logically complete and in strict correspondence with the psychological processes of thought.

In fact, no one engages in "logical reasoning" in the sense Reichenbach means, i.e. using first-order predicate calculus and modal logic. But we may be certain that "the scientific genius" he refers to does always feel bound to be rational in his scientific work! Reichenbach, however, understood logic to capture the essence of rationality; so when he systematized scientific method – which he took to be the location of scientific rationality – he assumed the system would be "logically complete" and he assumed that scientific method is value-free. For Reichenbach, good science is science carried out using good scientific methods that can be formulated using standard logic and which are value-free.

But those who do not assume the value-freedom of good scientific research, e.g. Giere, feminist philosophers of science, and others, are not using the phrase "good science" as it is used in science or in accordance with ordinary usage. When scientists refer to an interpretation of the data as "a good interpretation" or remark that a particular piece of work is "good science," they precisely do not mean that the interpretation is influenced by social or moral values as well as by whatever technical considerations the domain demands. Most people, *including scientists*, believe that when values influence scientific work in this way, the result is bad science.

To understand ordinary usage and scientific usage, we need to distinguish two senses of "good" as applied to scientific work. *Instrumentally good* scientific work is work leading to results that are positively evaluated using some category of epistemic success. Candidates include true or probably true, empirically successful, empirically adequate, and the like. Here Longino's general term, "conformation," is helpful. It names the category of ways in which representations, e.g. theories, hypotheses, models, graphs, etc., are related to objects in a domain. A representation might be isomorphic or homomorphic with the object(s), approximate or fit the object(s), be similar to or true of the object(s). Thus, representations etc. are successful if "they conform sufficiently to enable users to interact successfully with the domain of which they are theories or models"

(Longino 2002: 119). This will serve well as a characterization of what I am calling "instrumentally good scientific work." Their success might also, as many philosophers argue, depend upon their satisfying or maximizing some combination of constitutive values, but, as Miriam Solomon argues, their success might not depend on maximizing any constitutive values.

The philosophers treated in this book have argued that scientific work resulting in models, hypotheses, and theories can be successful work, can maximize constitutive values, and still depend upon, support, or maintain contextual values. If a particular theory depends upon (rests in part upon) and so maintains a benign moral value, we might say it is a *morally good* theory or decide that it is morally unobjectionable. In fact, if it maintains a common value, one that is part of the moral status quo, the theory is likely to pass as morally neutral.

I am arguing that the standard use of the phrase "good science" means instrumentally good science that is morally neutral. The value-neutrality thesis is embedded deeply enough in scientific and popular culture that calling a theory "good" because it maintains a widely accepted moral value is an *improper* sense of the phrase "good work." In ordinary usage and in common scientific parlance, "good science" means instrumentally good science that is morally value-neutral. Embedded in the ordinary usage (i.e. *non-philosophical*, standard usage) is the assumption that all good science is neutral among contextual values. And it follows from this assumption that scientific research influenced by any contextual values, good or bad, is instrumentally bad science.

Of course, only empirical investigation *that does not beg the question by assuming at the outset that good science is contextually value-neutral* can determine whether cases of instrumentally good science are or are not neutral among contextual values. And, indeed, studies that are methodologically agnostic about the contextual value-neutrality of instrumentally good science have produced plausible case histories revealing the influence of gender values and other contextual values upon instrumentally good science. Thus, the claim that any science influenced by contextual values is instrumentally bad science is either empirically premature or question begging.

Nevertheless, the value-neutrality thesis is as deeply embedded in the culture of *philosophy of science* as it is in science and in popular culture. Moral neutrality is not only a norm of good science but also a metaphilosophical norm governing good philosophy of science (although it usually functions tacitly). That is, traditional philosophies of science have agreed with scientists that (SN) good science is value-neutral and have built that norm into their philosophical accounts of scientific rationality; in the traditional view, a good philosophical account of scientific rationality shows that rational science is science in which the context of justification is free of contextual values. For traditional philosophers of science, (SN) means that (MPN 1) a *good philosophy of science* is set up to analyze and justify the morally neutral production of scientific knowledge. (MPN 1) is a metaphilosophical norm governing philosophy of science itself.

But most philosophies of science, *whether or not* they are traditional, i.e. embrace the scientific norm (SN) and metaphilosophical norm (MPN 1), nevertheless still (MPN 2) take contextual value-neutrality to be a norm governing good philosophy of science. (MPN 2) is a second metaphilosophical norm. Here we have distinguished a norm scientists agree upon: good scientific research should be morally neutral, and this same norm as it is taken up by philosophies of science and used to mark good, rational scientific research. The norm has been central to science and philosophy of science, directing philosophy of science to be blind to moral and political influences on good research. On the empirical side, philosophers are sometimes misled by case studies written under the influence of the value-neutrality norm of good science, for these are not likely to mention any contextual values at work in the context of justification of the cases. And even when case studies mention them, the norm (MPN 1) leads philosophers to rationally reconstruct the case to fit their own norms of good, rational scientific research and explicitly or tacitly make use of Reichenbach's apparatus, i.e. relegating that bit of history to the context of discovery, to external accounts of the case, and to the sociology or social history of science.

We have also distinguished a second metaphilosophical norm, viz. (MPN 2) philosophy of science should be neutral among moral, political, and other contextual values. Ought philosophies of science adopt this norm? Feminism is avowedly a standpoint with norms for moral and political engagement; thus, feminist philosophy of science is not thoroughly neutral among contextual values. Are critics right in arguing that this is necessarily bad philosophy of science because it is avowedly motivated by feminist values, argues that gender values influence instrumentally good science – not just in the choice of research projects and the uses to which results are put, but in the interpretation of data, testing of hypotheses, models, theories, and so on? It also argues that feminist values can help produce instrumentally better scientific research. Feminists, including feminist philosophers of science, do prefer science projects, models, hypotheses, and theories that, in addition to being instrumentally good, accord with, or at least do not violate, feminist values. By traditional lights, feminist philosophy of science is scandalous!

But we can treat philosophies of science as science scholars treat the sciences, i.e. we can analyze them to see whether in fact they are neutral among contextual values and, if not, we can find the locations at which values influence them. Here we will suggest two locations for the influence of contextual values upon any philosophy of science:

in its descriptive aspects, including highly rationally reconstructed examples, and

in its choices among metaphilosophical norms.

6.1 Descriptive aspects

To illustrate or support their philosophies of science, philosophers of science appeal to case studies of work in the sciences, occasionally do case studies, and frequently rationally reconstruct cases. In doing a history or sociology of science or rationally reconstructing a case, a philosopher must, like any historian or sociologist, decide which facts are relevant and which are significant and should be included and explained. In appealing to a case study, the philosopher endorses its selection of facts as an adequate representation of the actual case. But how does the philosopher decide when an account, his own or someone else's, is representationally adequate? For example, when can a philosophical account ignore *indications* that a group of scientists were as a matter of fact influenced by a contextual value in the interpretation of their data? The historical record is almost never crystal clear so let us suppose just such a historical account was published by a young feminist historian in the journal, *Social Studies of Science*. Is a philosophy of science descriptively and explanatorily adequate if it ignores the indications put forward in the account? Or suppose the philosopher does his own primary research on one of the paradigm cases in the history of science and finds *indications* that, in their instrumentally adequate interpretation of the experimental data, scientists were influenced by political considerations. How strong must the indications be to require philosophical notice? When can the philosopher ignore them?

These are important questions unless, as we noted above, the philosopher has merely begged the question at issue by assuming that good science is neutral among contextual values. In making his decision, the philosopher of science can be influenced by the desiderata of good history or sociology of science, by philosophical desiderata, by professional interests, by social and political interests, or by background assumptions that include socially or politically significant interests or values.

In each case, we can only determine whether and how such factors influenced the philosopher's decision by empirical investigation.

6.2 Metaphilosophical norms

Every philosopher of science chooses among norms governing philosophy of science. A good philosophy of science should maximize some combination of what we might call "philosophical virtues," those norms contributing to the success of any philosophy, as well as norms contributing to its success as a philosophy of science. A first pass suggests this (hopefully uncontroversial) list of philosophical virtues: precision, thoroughness, consistency, inferential (whether deductive, inductive, abductive, or analogical) rigor, making as many relevant distinctions as possible, and being philosophically interesting – although this may be a professional value rather than a virtue of philosophies. I am not convinced that philosophies need to be sound since it is hard to come by true yet interesting premises, and it is difficult to imagine putting a philosophy into proper logical form. This list is suggestive, not conclusive; there are undoubtedly other general philosophical norms.

Turning to philosophy of science, we are most concerned with norm (MPN 1) a good philosophy of science takes instrumentally good science to be neutral among contextual values and is set up to (MPN 1.1) analyze and (MPN 1.2) justify the value-neutral production of scientific knowledge, and norm (MPN 2) a good philosophy of science is neutral among contextual values. (Contrary norms include MPN 1', a good philosophy of science shows that and how instrumentally good science can be influenced by contextual values, and MPN 2', a good philosophy of science need not be neutral among contextual values.)

Suppose we are working on our own philosophy of science and we assume as metaphilosophical norms that (MPN 1) our philosophy of science should assume the value-neutrality of science and (MPN 2) our philosophy of science should be neutral among contextual values. Following MPN 1, if contextual values are found to influence the methods or results of instrumentally good research, our philosophy of science must either (1) label it "bad science" – despite scientists themselves treating it as good work, or find a way to be, as it were, blind to the information, e.g. (2) by relegating indications of such influence to the context of discovery or ignoring the information altogether by (3) rationally reconstructing the case. Strategy (1) renders our philosophical account unconvincing; but if we adopt strategy (2) or (3), our philosophy of science acquiesces in, tacitly accepts, the moral, social, political, or economic arrangements maintained by the actual research in question. It does so because any scientific work depending upon contextual values in testing hypotheses, interpreting data, and the like thereby maintains those values.

Although strategies (2) and (3) *appear* to accept only the instrumental success of the work without endorsing the values it historically depends upon or supports, they do not succeed. Adopting strategies (2) and (3) is "betting" that the hypothesis can be, *without these values*, proven in some way acceptable to scientists working in the relevant domain. But, to date, philosophers have not undertaken the job of reworking more than a meager handful of hard-fought cases. And few of these "reworked" cases have been vetted by the relevant scientific community let alone endorsed by it. Thus, the "bet" that most cases can be reworked in this way comes down to faith (which is not something philosophers are fond of indulging professionally).

In the absence of such "reworked" cases, by accepting the actual cases as instrumentally good work, our imaginary philosopher of science is stuck with the contextual values in fact maintained by the work. But this clearly violates the second metaphilosophical norm we adopted, (MPN 2) a philosophy of science should be neutral among contextual values.

Moreover, whether we adopt MPN 1 and 2 or MPN 1' and 2', if we deem any scientific research depending upon or supporting contextual values instrumentally good work, we explicitly or implicitly, directly or indirectly, endorse those values in whatever way the scientific work maintains them. This is so whether or not there exists a case study revealing the influence of contextual values upon the methods or results of a research project. Nor is our argument defeated by the existence of cases in which the research really is free of the influence of contextual values – in which case, we do not endorse any contextual values. But our argument does show that we can no longer just assume that philosophies of science are neutral among contextual values.

This is especially so for philosophers of science if they do case studies or rationally reconstruct cases. As Anderson argues of scientists (see Chapter 3), philosophers, too, need interests to guide their empirical research as they determine which facts are significant. Our guiding interests might be limited to technical ones. But to the degree that a philosophy of science aims to be useful to our lives, to that degree the guiding interests are likely to be contextual ones. (This is a rule of thumb for feminist philosophers of science.)

Feminist philosophies of science wear their central contextual value on their philosophical sleeves. The burden of the argument in this chapter has been to suggest that most other philosophies of science are "trying to pass" as neutral among contextual values.

As we have seen in the pages of this book, admitting contextual values to the purview of philosophy of science changes our understanding of science and our self-understanding as philosophers of science. While it is not irrational for scientists to use contextual values in making decisions among competing hypotheses etc., it is irrational for them to do so un-self-consciously, without careful thought. Thus, as some of the philosophers we have considered argue, it is irrational not to build self-reflection into scientific method. In the same way, it is not irrational for philosophers to use contextual values in producing or supporting their philosophical accounts. But it is irrational to do so without careful thought or without any awareness of doing so.

NOTES

INTRODUCTION

1 "A theory of X-linkage of major intellectual traits," *American Journal of Mental Deficiency* 76 (1972): 611–19. Cited in Fausto-Sterling 1985: 18. Fausto-Sterling sets out enough background to enable non-science readers to understand Lehrke's arguments in some detail. She also mentions cogent criticisms of the variability hypothesis and offers her criticism of the fundamental assumption that "intelligence is an inherited trait coded for by some finite number of factors called 'intelligence genes'" (Fausto-Sterling 1985: 19–20).

4 FEMINIST CONTEXTUAL EMPIRICISM

1 We should note that Longino's notion of "doxastic autonomy" is very different from the notion of "doxastic autonomy" we have attributed to Nelson. Nelson argues that individuals cannot have beliefs autonomously, meaning in epistemic isolation, from others in her community. Longino would not deny this; her point is that individuals can differ from one another in their beliefs.

5 STANDPOINT EPISTEMOLOGIES OF SCIENCE

1 In (1998), Collins refines her understanding of a standpoint and, with it, her notion of the outsider within. Nevertheless, her definition of "standpoint theory" is still quite compatible with Harding's:

> a social theory arguing that group location in hierarchical power relations produces shared experiences for individuals in those groups, and that these common experiences can foster similar angles of vision leading to a group knowledge or standpoint deemed essential for informed political action (p. 281).

She does not offer a new definition of outsider within, but notes that such a location enables "remarkable critical social theory" (p. 231). Moreover, those who are critical of their own privilege can move into outsider within spaces (p. 234).

While Collins does not claim that outsiders within have epistemic advantage over everyone else, she maintains that truth is constructed, "determined," or "negotiated" (pp. 237, 231, 238, 232) in the first instance by those in outsider within locations. Responding to the question whether truth can be dialogically determined by groups with competing interests, she says that black women outsiders within are confrontational in response to different interests and collaborative in response to shared interests (p. 238). This suggests that those with different standpoints might agree about what is true when they have shared interests and disagree when they do not.

6 CONCLUDING ARGUMENTS: CAN THE PHILOSOPHY OF SCIENCE BE VALUE-FREE?

1 This despite Hesse's anti-feminist stance. We may note that Hesse advocates a postmodern philosophy of science, rejecting the Ontological Tyranny and recognizing the local nature of knowledge claims, yet, on her reading, feminist epistemologies and philosophies of science are untenable (see Hesse 1994).

BIBLIOGRAPHY

Addelson, Kathryn P. (1983) "The man of professional wisdom," in Sandra
 Harding and Merrill Hintikka (eds) Discovering Reality: Feminist Perspectives on Epistemology, Metaphysics, Methodology, and Philosophy of Science, Dordrecht: D. Reidel, pp.
 165–86.

Alcoff, Linda and Potter, Elizabeth (eds) (1993) Feminist Epistemologies, New York and
 London: Routledge.

Amann, Klaus and Knorr-Cetina, Karin (1990) "The fixation of (visual) evidence,"
 in Michael Lynch and Steve Woolgar (eds) Representation in Scientific Practice,
 Cambridge, MA: MIT Press, pp. 85–122.

Anderson, Elizabeth (2004) "Uses of value judgments in science: a general argument, with lessons from a case study of feminist research on divorce," Hypatia
 19(1): 1–24.

—— (1995a) "Feminist epistemology: an interpretation and a defense," Hypatia
 10(3): 50–84.

—— (1995b) "Knowledge, human interests, and objectivity in feminist epistemology," Philosophical Topics 23(2): 27–58.

Bakhurst, David (1994) "Ideology," in J. Dancy and E. Sosa (eds) A Companion to
 Epistemology, Oxford, UK, and Cambridge, MA: Blackwell, pp. 191–3.

Bleier, Ruth (ed.) (1988) Feminist Approaches to Science, New York: Pergamon Press.

—— (1984) Science and Gender, Elmsford, NY: Pergamon.

Brumfiel, E.M. (1991) "Weaving and cooking: women's production in Aztec
 Mexico," in J.M. Gero and M.W. Conkey (eds) Engendering Archaeology: Women and
 Prehistory, Oxford: Blackwell, pp. 224–53.

Burian, Richard M. (1993) "Technique, task definition, and the transition from
 genetics to molecular genetics," Journal of the History of Biology 26(3): 387–407.

—— (1985) "The 'internal politics' of biology and the justification of biological
 theories," in A. Donagan, N. Perovich, and M. Wedin (eds) Human Nature and
 Natural Knowledge, Dordrecht: D. Reidel.

Burtt, E.A. (1932) The Metaphysical Foundations of Modern Physical Science, Garden City, NY:
 Doubleday/Anchor.

Chi, J.G., Dooling, E.C., and Gilles, F.H. (1977) "Gyral development of the human
 brain," Annals of Neurology 1.

Chodorow, Nancy (1978) *The Reproduction of Mothering*, Berkeley: University of California Press.

Collins, Patricia Hill (1998) *Fighting Words: Black Women and the Search for Justice*, Minneapolis: University of Minnesota Press.

—— (1991) *Black Feminist Thought: Knowledge, Consciousness, and the Politics of Empowerment*, New York: Routledge.

—— (1986) "Learning from the outsider within: the sociological significance of black feminist thought," *Social Problems* 33: 35–9.

Conant, James Bryant (1970) *Robert Boyle's Experiments in Pneumatics*, Cambridge, MA: Harvard University Press.

Deigh, John (1994) "Cognitivism in the theory of emotions," *Ethics* 104: 824–54.

Diamond, M.C., Dowling, G., and Johnson, R. (1981) "Morphologic cerebral cortical asymmetry in male and female rats," *Experimental Neurology* 71.

Downes, S.M. (1993) "Socializing naturalized philosophy of science," *Philosophy of Science* 60: 452–68.

Duhem, Pierre (1954) *The Aim and Structure of Physical Theory*, trans. Philip P. Wiener, Princeton, NJ: Princeton University Press.

Dupré, John (1996) "Metaphysical disorder and scientific disunity," in Peter Galison and David J. Stump (eds) *The Disunity of Science*, Stanford: Stanford University Press, pp. 101–17.

—— (1993) *The Disorder of Things*, Cambridge, MA: Harvard University Press.

Ehrhardt, Anke and Meyer-Bahlburg, Heino (1981) "Effects of prenatal sex hormones on gender-related behavior," *Science* 211: 1312–18.

Fausto-Sterling, Anne (2004) "The five sexes: why male and female are not enough," in Lisa Heldke and Peg O'Connor (eds) *Oppression, Privilege and Resistance: Theoretical Perspectives on Racism, Sexism, and Heterosexism*, New York: McGraw-Hill, pp. 476–82. Originally published in *The Sciences* (March/April 1993).

—— (2000) *Sexing the Body: Gender Politics and the Construction of Sexuality*, New York: Basic Books, Inc.

—— (1985) *Myths of Gender*, New York: Basic Books, Inc.

Flax, Jane (1986) "Gender as a social problem: in and for feminist theory," *American Studies/Amerika Studien*, journal of the German Association for American Studies.

Fuller, Steve (1988) *Social Epistemology*, Bloomington and Indianapolis: Indiana University Press.

Galison, Peter (1996) "Computer simulations and the trading zone," in Peter Galison and David J. Stump (eds) *The Disunity of Science*, Stanford, CA: Stanford University Press, pp. 118–57.

Geertz, Clifford (1990) "A lab of one's own," *New York Review of Books* 37 (8 November).

—— (1976) "From the native's point of view: on the nature of anthropological understanding," in K.H. Basso and H.A. Selby (eds) *Meaning in Anthropology*, School of American Research Advanced Seminar Series, Albuquerque: University of New Mexico Press, pp. 221–37.

Gero, J. M. (1993) "The social world of prehistoric facts: gender and power in Paleoindian research," in H. duCros and L. Smith (eds) *Women in Archaeology: A Feminist Critique*, Research School of Pacific Studies, Occasional Papers in Prehistory, no. 23, Canberra: Australian National University, pp. 31–40.

Geschwind, N. and Behan, P. (1984) "Laterality, hormones, and immunity," in N. Geschwind and A.M. Galaburda (eds) *Cerebral Dominance: The Biological Foundations*, Cambridge, MA: Harvard University Press, pp. 211–24.

—— (1982) "Left-handedness: association with immune disease, migraine, and developmental learning disorder," *Proceedings of National Academy of Sciences 79*.

Geschwind, N. and Galaburda, A.M. (1984) *Cerebral Dominance: The Biological Foundations* Cambridge, MA: Harvard University Press.

Giere, Ronald N. (1988) *Explaining Science*, Chicago and London: University of Chicago Press.

Gorski, R. (1979) "The neuroendocrinology of reproduction: an overview," *Biology of Reproduction 20*.

Gorski, R., Gordon, J.H., Shryne, J.E., and Southam, A.M. (1978) "Evidence for a morphological sex difference within the meidal preoptic area of the rat brain," *Brain Research 148*.

Gould, Steven J. (1981) *The Mismeasure of Man*, New York: W.W. Norton.

Gross, Paul and Levitt, Norman (1994) *Higher Superstition*, Baltimore: Johns Hopkins University Press.

Haack, Susan (1993) "Epistemological reflections of an old feminist," *Reason Papers* 18: 31–43.

Haraway, Donna (1991a) "Situated knowledges: the science question in feminism and the privilege of partial perspective," in *Simians, Cyborgs, and Women: The Reinvention of Nature*, New York: Routledge, pp. 183–203.

—— (1991b) *Simians, Cyborgs and Women*, New York: Routledge.

—— (1989) *Primate Visions*, New York: Routledge.

Hardin, Margaret F. (1970) "Design structure and social interaction: archaeological implications of an ethnographic analysis," *American Antiquity* 35: 332–43.

Harding, Sandra (1998) *Is Science Multicultural? Postcolonialisms, Feminisms, and Epistemologies*, Bloomington and Indianapolis: Indiana University Press.

—— (1997) "Comment on Hekman's 'truth and method: feminist standpoint theory revisited': Whose standpoint needs the regimes of truth and reality?" *Signs: Journal of Women in Culture and Society* 22(2): 382–91.

—— (1993) "Rethinking standpoint epistemology: what is 'strong objectivity'?", in Linda Alcoff and Elizabeth Potter (eds) *Feminist Epistemologies*, New York and London: Routledge, pp. 49–82.

—— (1991) *Whose Science? Whose Knowledge? Thinking From Women's Lives*, Ithaca: Cornell University Press.

—— (1986) *The Science Question in Feminism*, Ithaca: Cornell University Press.

—— (1983) "Why has the sex/gender system become visible only now," in Sandra Harding and Merrill Hintikka (eds) *Discovering Reality: Feminist Perspectives* on *Epistemology, Metaphysics, Methodology, and Philosophy of Science*, Dordrecht: D. Reidel, pp. 311–24.

Harris, G., and Levine, S. (1965) "Sexual differentiation of the brain and its experimental control," *Journal of Physiology* 181.

Hartsock, Nancy (1983) "The feminist standpoint: developing the ground for a specifically feminist historical materialism," in Sandra Harding and Merrill Hintikka (eds) *Discovering Reality: Feminist Perspectives on Epistemology, Metaphysics, Methodology, and Philosophy of Science*, Dordrecht: Reidel/Kluwer, pp. 283–310.

Hastorf, C.A. (1991) "Gender, space, and food in prehistory," in J.M. Gero and M.W. Conkey (eds) *Engendering Archaeology: Women and Prehistory*, Oxford: Blackwell, pp. 132–59.

Hayles, N. Katherine (1993) "Constrained constructivism: locating scientific inquiry in the theater of representation," in George Levine (ed.) *Realism and Representation*, Madison: University of Wisconsin Press, pp. 27–43.

Hekman, Susan (1997) "Truth and method: feminist standpoint theory revisited," *Signs* 22(2): 341–402.

Hempel, C.G. (1981) "Turns in the evolution of the problem of induction," *Synthese* 46: 389–404.

Hesse, Mary (1994) "How to be postmodern without being a feminist," *The Monist* 77(4): 445–61.

—— (1974) *The Structure of Scientific Inference*, Berkeley and Los Angeles: University of California Press.

Hill, J.N. (1966) "A prehistoric community in eastern Arizona," *Southwestern Journal of Anthropology* 22: 9–30.

Hubbard, Ruth (1990) *The Politics of Women's Biology*, New Brunswick, NJ: Rutgers University Press.

—— (1982) "Have only men evolved?" in Ruth Hubbard, M. Henifin, and B. Fried (eds) *Biological Woman – The Convenient Myth*, Cambridge, MA: Schenkman.

Hyde, Janet S. (1981) "How large are cognitive differences? A meta-analysis using ω and d," *American Psychologist* 36: 892–901.

Jaggar, Alison (1983) *Feminist Politics and Human Nature*, Totowa, NJ: Rowman & Allenheld.

Jordanova, Ludmilla (1993) *Sexual Visions: Images of Gender in Science and Medicine between the Eighteenth and Twentieth Centuries*, Madison: University of Wisconsin Press.

Keller, Evelyn Fox (1992) *Secrets of Life, Secrets of Death: Essays on Language, Gender, and Science*, New York: Routledge.

—— (1985) *Reflections on Gender and Science*, New Haven, CT: Yale University Press.

Kitcher, Philip (2001) *Science, Truth and Democracy*, Oxford and New York: Oxford University Press.

—— (1993) *The Advancement of Science: Science without Legend, Objectivity without Illusions*, Oxford: Oxford University Press.

Knorr-Cetina, Karin (1981) *The Manufacture of Knowledge*, Oxford: Pergamon Press.

Kuhn, Thomas S. (1970) *The Structure of Scientific Revolutions*, Chicago: The University of Chicago Press.

Lacey, Hugh (1999) *Is Science Value Free? Values and Scientific Understanding*, New York: Routledge.

Lewontin, Richard, Rose, Steven, and Kamin, Leon (1984) *Not in Our Genes*, New York: Pantheon.

Lloyd, Elisabeth (1995) "Objectivity and the double standard for feminist epistemologies," *Synthese* 104: 351–81.

Longacre, W.A. (1968) "Some aspects of prehistoric society in east-central Arizona," in S.R. Binford and L.R. Binford (eds) *New Perspectives in Archeology*, Chicago: Aldine, pp. 89–102.

—— (1966) "Changing patterns of social integration: a prehistoric example for the American Southwest," *American Anthropologist* 68: 94–102.

Longino, Helen (2002) *The Fate of Knowledge*, Princeton, NJ: Princeton University Press.

—— (1994) "In search of feminist epistemology," *The Monist* 77(4): 472–85.

—— (1990) *Science as Social Knowledge*, Princeton, NJ: Princeton University Press.

Maccoby, Eleanor and Jacklin, Carol (1974) *The Psychology of Sex Differences*, Stanford, CA: Stanford University Press.

Maffie, James (1991) "What is social about social epistemics?" *Social Epistemology* 5: 101–10.

Narayan, Uma (1988) "Working together across difference: some considerations on emotions and political practice," *Hypatia* 3(2): 31–48.

Nelson, Lynn Hankinson (2000) *On Quine*, Belmont, CA: Wadsworth/Thomson.

—— (1996) "Empiricism without dogmas," in Lynn Hankinson Nelson and Jack Nelson (eds) *Feminism, Science and the Philosophy of Science*, Dordrecht, Boston, and London: Kluwer Academic Publishers, pp. 95–119.

—— (1995) "Feminist naturalized philosophy of science," *Synthese* 104(3): 39 –421.

—— (1993) "Epistemological communities," in Linda Alcoff and Elizabeth Potter (eds) *Feminist Epistemologies*, New York and London: Routledge, pp. 121–59.

—— (1990) *Who Knows: From Quine to Feminist Empiricism*, Philadelphia: Temple University Press.

Nelson, Lynn Hankinson and Nelson, Jack (eds) (2003) *Feminist Interpretations of W.V. Quine*, University Park: Pennsylvania State University Press.

—— (eds) (1996) *Feminism, Science and the Philosophy of Science*, Dordrecht, Boston, and London: Kluwer Academic Publishers.

—— (1994) "No rush to judgment," *The Monist* 77(4): 486–508.

Paul, Diane (1995) *Controlling Human Heredity*, New York: Humanities Press.

Potter, Elizabeth (2001) *Gender and Boyle's Law of Gases*, Bloomington and Indianapolis: Indiana University Press.

—— (1995) "Good science and good philosophy of science," *Synthese* 104(3): 423–39.

—— (1993) "Gender and epistemic negotiation," in Linda Alcoff and Elizabeth Potter (eds) *Feminist Epistemologies*, New York and London: Routledge, pp. 161–86.

—— (1988) "Modeling the gender politics in science," *Hypatia: A Journal of Feminist Philosophy* 3(3). Reprinted in *Feminism and Science*, ed. Nancy Tuana, Bloomington and Indiana: Indiana University Press, 1989.

Proctor, Robert (1988) *Racial Hygiene: Medicine under the Nazis*, Cambridge, MA: Harvard University Press.

Putnam, Hilary (1983) "Why reason can't be naturalized," in *Realism and Reason*, New York: Cambridge University Press, pp. 229–47.

Quine, W.V. and Ullian, J.S. (1970) *The Web of Belief*, New York: Random House.

Reichenbach, Hans (1938) *Experience and Prediction*, Chicago: University of Chicago Press.

Searle, J.R. (1992) *The Rediscovery of the Mind*, Cambridge, MA: MIT Press.

—— (1984) *Minds, Brains and Science*, Cambridge, MA: Harvard University Press.

Smith, Dorothy (1990) *The Conceptual Practices of Power: A Feminist Sociology of Knowledge*, Boston: Northeastern University Press.

—— (1987) *The Everyday World as Problematic: A Sociology for Women*, Boston: Northeastern University Press.

Solomon, Miriam (2001) *Social Empiricism*, Cambridge, MA: Massachusetts Institute of Technology.

—— (1994) "A more social empiricism," in Fred Schmitt (ed.) *Socializing Epistemology: The Social Dimensions of Knowledge*, Lanham, MD: Rowan & Littlefield, pp. 217–33.

Spanier, Bonnie (1995) *Im/partial Science: Gender Ideology in Molecular Biology*, Bloomington and Indianapolis: Indiana University Press.

Spelman, Elizabeth V. (1988) *Inessential Woman: Problems of Exclusion in Feminist Thought*, Boston: Beacon Press.

Stewart, Abigail J., Copeland, Anne P., Chester, Nia Lane, Malley, Janet E., and Barenbaum, Nicole B. (1997) *Separating Together*, New York: Guilford Press.

Stump, David (1992) "Naturalized philosophy of science with a plurality of methods," *Philosophy of Science* 59: 456–60.

Tiles, Mary (1987) "A science of Mars or of Venus?" *Philosophy* 62: 293–306.

Van Fraassen, Bas C. (1980) *The Scientific Image*, Oxford: Clarendon Press.

Walker, B.M. (1981) "Psychology and feminism – if you can't beat them, join them," in Dale Spender (ed.) *Men's Studies Modified*, New York, Pergamon Press, pp. 111–24.

Wallerstein, J.S. and Kelly, J.B. (1980) *Surviving the Breakup*, New York: Basic.

Wallerstein, Judith, Lewis, Julia, and Blakeslee, Sandra (2000) *The Unexpected Legacy of Divorce*, New York: Hyperion.

Watson, Patty Jo and Kennedy, M.C. (1991) "The development of horticulture in the eastern woodlands of North Am: women's role," in J.M. Gero and M.W. Conkey (eds) *Engendering Archaeology: Women and Prehistory*, Oxford: Blackwell, pp. 255–75.

Weber, Max (1946) "Science as a vocation," in H.H. Gerth and C. Wright Mills (eds) *From Max Weber: Essays in Sociology*, New York: Oxford University Press.

Wylie, Alison (2003) "Why standpoint matters," in Robert Figueroa and Sandra Harding (eds) *Science and Other Cultures: Issues in Philosophies of Science and Technology*, New York: Routledge, pp. 26–48.

—— (2002) *Thinking from Things: Essays in the Philosophy of Archaeology*, Berkeley: University of California Press.

—— (1997) "The engendering of archaeology: refiguring feminist science studies," *Osiris* 12: 80–99.

—— (1994) "Doing philosophy as a feminist: Longino on the search for a feminist epistemology," *The Monist* 77(4): 345–58.

Zimmerman, Michael J. (2003) "Intrinsic vs. extrinsic value," in Edward N. Zalta (ed.) *The Stanford Encyclopedia of Philosophy* (fall 2003 edition), http://plato.stanford.edu/archives/fall2003/entries/value-intrinsic-extrinsic/.

INDEX

Printed in the United States
by Baker & Taylor Publisher Services